More Evidence Against the Random Walk Hypothesis

More Evidence Against the Random Walk Hypothesis

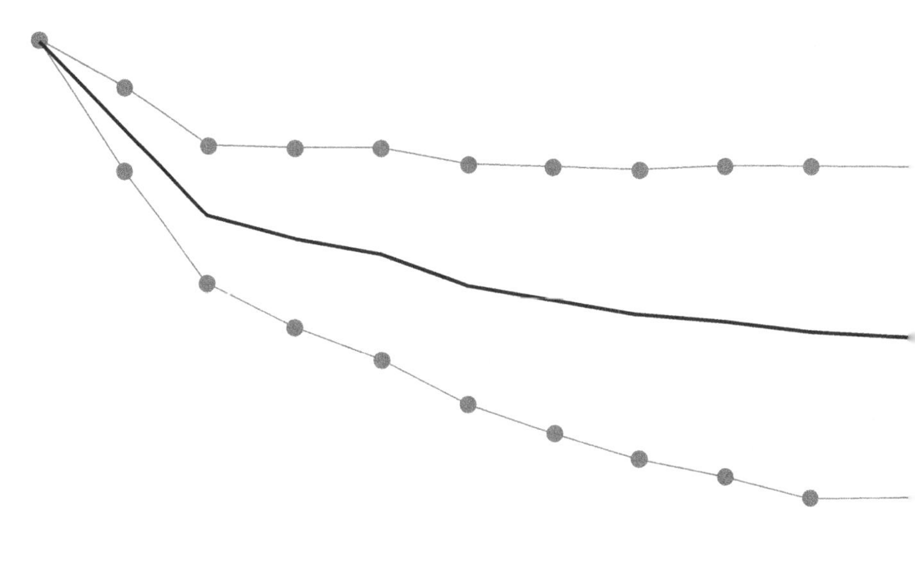

Shunxin Jiang

World Scientific

NEW JERSEY • LONDON • SINGAPORE • BEIJING • SHANGHAI • HONG KONG • TAIPEI • CHENNAI

Published by

World Scientific Publishing Co. Pte. Ltd.

5 Toh Tuck Link, Singapore 596224

USA office: 27 Warren Street, Suite 401-402, Hackensack, NJ 07601

UK office: 57 Shelton Street, Covent Garden, London WC2H 9HE

Library of Congress Cataloging-in-Publication Data
Jiang, Shunxin.
More evidence against the random walk hypothesis / Shunxin Jiang.
pages cm
Includes bibliographical references and index.
ISBN 978-981-4641-05-0 (hardcover : alk. paper)
1. Exchange traded funds. I. Title.
HG6043.J53 2015
332.63'27--dc23
2014037033

British Library Cataloguing-in-Publication Data
A catalogue record for this book is available from the British Library.

Printed in Singapore

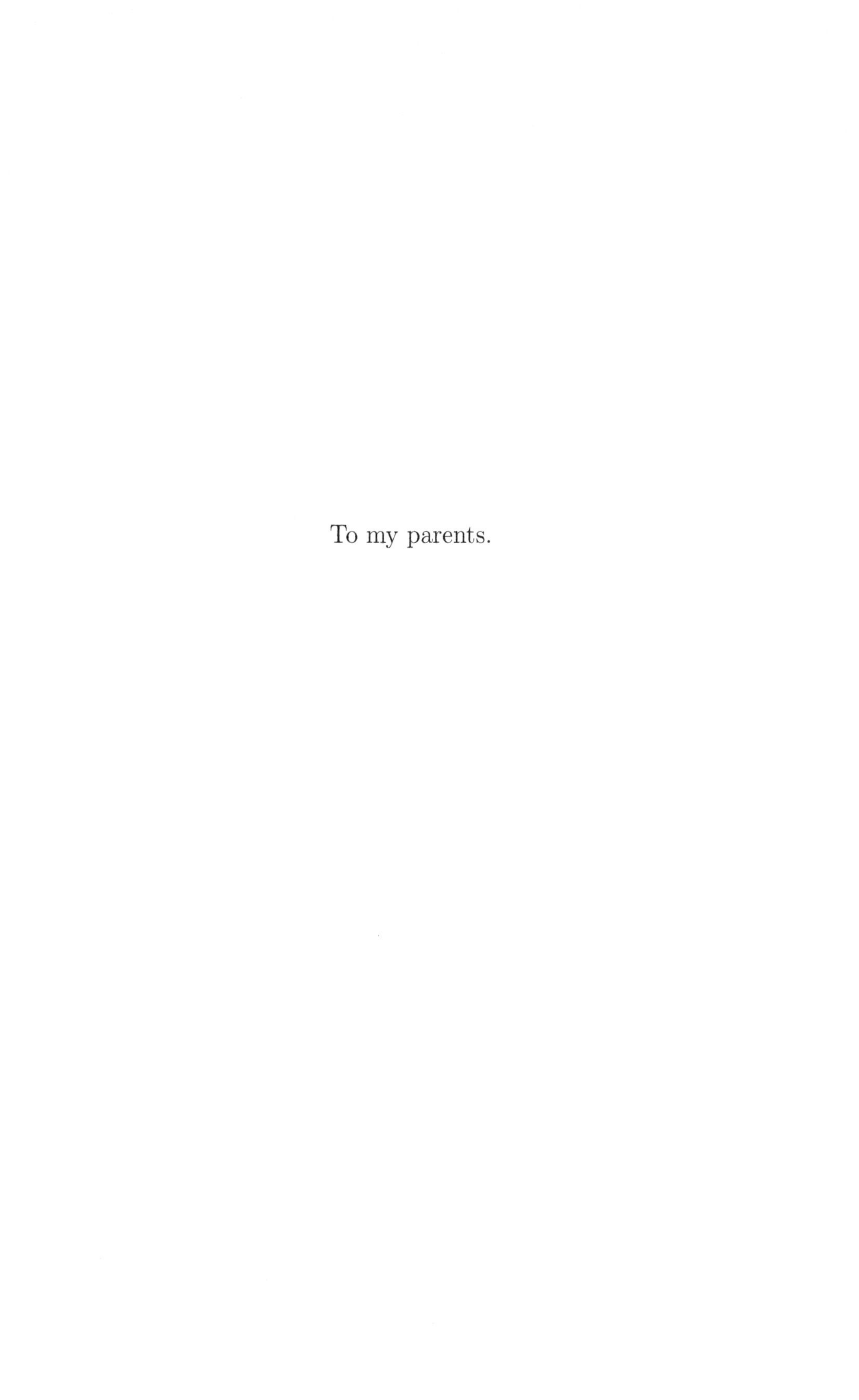

To my parents.

Preface

This volume grew out of my thesis research on the exchange traded fund (ETF) market and volatility trading while I was a doctoral student in mathematics at the Courant Institute from 2010 to 2014. It provides more evidence against the Random Walk Hypothesis and offers insights into market inefficiency through systematically trading ETFs. The book is organized to answer the following three questions: (1) Do ETF prices follow random walks? (2) If not, what are some of the factors that impact their non-random walk behavior? (3) How can we take advantage of such price dynamics in trading ETFs? As to whether this material will lead to greater wealth, I can only hope to find out from practice.

During the course of my graduate research, I have accumulated great debts to my advisor, Marco Avellaneda. I am deeply grateful to his constant support and guidance. His immense knowledge in quantitative finance and unique creativity in analyzing complex problems have tremendously influenced me. This work would not have been possible without his supervision.

I would also like to express my sincere gratitude to Robert Kohn and Jonathan Goodman for offering many constructive comments and valuable suggestions for improvement. I wish to thank all those who have contributed to my growth over my years at the Courant Institute - Gerard Ben Arous, Maria Cameron, Petter Kolm, Henry McKean, Charles Newman, Esteban Tabak, S.R. Srinivasa Varadhan, Olof Widlund, Lai-Sang Young.

I gratefully acknowledge the funding sources which allowed me to pursue my doctoral studies: the MacCracken Fellowship Program, Frank J. Gould

Chair Fund Fellowship, and Jeffrey and Denise Rosenbluth Fellowship.

My time at Courant was made enjoyable in large part due to the many friends that became a part of my life, I feel very fortunate to have them with me through difficult times.

Lastly, I would like to thank my family for their love and support. They are everything to me; they made me believe that anything is possible.

Shunxin Jiang
August 30, 2014

Contents

List of Figures

List of Tables

Chapter 1

ETF Prices Do Not Follow Random Walks

1.1 Introduction

Fama (1970) first theorized the notion of capital market efficiency. He argued that a market in which prices at any time always fully reflect all available information should be called efficient. Based on the specification of the term "all available information" in a given capital market, Fama further defined three particular forms of efficiency: the weak form, the semi-strong form and the strong form. The *weak form* efficiency deals with the information set which contains only the market trading data such as historical prices and trading volumes. The *semi-strong form* efficiency extends the information set of the weak form by including other information that is publicly available such as announcements of earnings, stock splits and economic data, etc. The *strong form* efficiency is concerned with whether market participants have monopolistic access to all information relevant for asset price formation, whether it be public or private. In essence, under Fama's market efficiency hypothesis, investors will not be able to systematically profit from any deviation of asset prices from their equilibrium levels, though such deviations may exist.

The efficiency of a capital market has important implications for its participants and the economy as a whole. It allows firms to make effective production-and-investment decisions and households to allocate their income into saving, investment and consumption with more security and precision. Efficient markets tend to exhibit more liquidity, breadth and depth; they are usually fully deregulated and liberalized. In an efficient market, expected capital investment returns will reflect closely their risk profiles so that goods trade at their fair values. Economies with such capital markets tend to attract more exports and foreign investments. On the other hand, if capital markets are inefficient, more opportunities for speculations may occur. Certain price patterns can be more predictable so that technical analysis may capture distortions between risk and return, achieving excess returns. Information may not be distributed to investors evenly or in a timely manner so that one group of investors may front-run the others or insiders may take advantage of their private information achieving unfair profits.

From the early empirical research in the 1960s and 1970s, the efficient market hypothesis received wide support. For instance, Jensen (1967) investigated the performance of 115 mutual fund managers in the period 1945-1964 and concluded that these mutual funds were on average not able to predict security prices well enough to outperform the market, even under the assumption that their bookkeeping, research, and other expenses except brokerage commissions were obtained freely. As a supporter of the Efficient Market Hypothesis, he later wrote [Jensen (1978)] "I believe there is no other proposition in economics which has more solid empirical evidence supporting it". Fama, who is Jensen's academic advisor, in 1970 compiled even more evidences in his paper *Efficient Capital Markets: A Review of Theory and Empirical Work* to support the efficient market hypothesis. In the paper, he revisited various pieces of empirical research and discussed how stock prices efficiently adjust to new information during stock splits and public announcements (including company earning announcements, Federal Reserve discount rate announcements, etc.).

However, the quest to challenge this framework did not stop. Early treatments of the efficient market models mostly favored the random walk model, in which the logarithmic asset price, X_t, follows a stochastic process represented by the following relation

$$X_t = \mu + X_{t-1} + w_t, \tag{1.1}$$

where μ is an arbitrary drift parameter and w_t's are independent random variables.[1] This formulation is usually associated with the weak form efficiency and its suitability is intuitively justifiable. For instance, it makes

[1]The random walk model was first developed by Louis Bachelier in his doctoral thesis in 1900 and proposed to be the fundamental model for financial time series.

sense that when the prices already fully reflect all available information, their successive changes should be more or less independent from one another. If not, say, increases in prices tend to be followed by increases in prices, then whenever a jump in stock prices is observed, market participants will attempt to go long on the stock in the subsequent periods so that this pattern gets fully exploited. Hence, testing the log price series against the random walk behavior became one of the most widely-adopted ways to examine the efficiency of a given capital market.

Soon, however, researchers began to question whether a random walk, or more generally a martingale sequence, is the proper characterization of the log asset price within an efficient capital market. Some constructed explicit efficient markets, where martingale models may not necessarily hold. For instance, LeRoy (1973) found that asset prices need not form a martingale sequence, when the expected rate of return on a stock is explained in terms of the portfolio optimization of risk-averse investors rather than simply taken as given. Under general conditions such as risk-aversion, the martingale property will be satisfied only as an approximation. The testing of market efficiency, therefore, should be model-specific and a good understanding of the return-generating process of asset prices is indispensable for this purpose.

Essentially, this leads to the joint hypothesis problem. When testing market efficiency, one needs to answer at least two basic questions: (1) what does an efficient market look like? (2) how do we test it statistically? The former question typically demands an equilibrium pricing model which incorporates investors' preferences and information structure; whereas, the latter involves hypothesis testing. Based on the joint hypothesis formulation, a test of the efficient market naturally turns into a test of multiple auxiliary hypotheses. A rejection here would in fact tell us very little about which aspect of problem is inconsistent with the data. For instance, a rejection of the joint hypothesis problem may suggest that the market is inefficient, that our description of the efficient market is incorrect, or both. Therefore, testing market efficiency is problematic, if not impossible. The materials in this volume can be considered as a slightly different approach in understanding market efficiency. We first test asset prices against the random walk hypothesis and then design systematic strategies which will profit from the non-random walk price behavior. By successfully doing so, we practically provide additional evidence that markets are not that efficient. What is in need now is a robust method which tests against the random walk model.

In 1988, Lo and MacKinlay (1988) developed probably one of the most successful specification tests to refute the random walk hypothesis on stock logarithmic prices. The test is based on the ratio of variance estimators of stock returns at different frequencies and is robust under heteroscedasticity and non-normality. Using this variance ratio statistic, they provide solid evidence that stocks do not follow random walks. Their empirical results strongly reject the random walk hypothesis for the entire sample period 1962-1985 and for all sub-periods on various aggregate return indexes and size-sorted portfolios. The variance ratio test they proposed has since been widely adopted by an ever-increasing number of studies [Lee *et al.* (2001); Cheung and Coutts (2001); Dezelan (2000); Grieb and Reyes (1999); Groenewold and Ariff (2002a,b); Huang (1995); Poon (1996); Worthington and Higgs (2004)] in understanding market efficiency. A majority of the authors have chosen to focus on either the stock market or the foreign exchange market of individual countries or regions. Few have investigated the efficiency of the exchange traded fund (ETF) market.

One of the reasons why ETFs are missed is probably that the ETF market is rather young. Its very first product[2] in the United States was created in January 1993, known as the SPDR S&P 500 ETF, which like an index fund represents the basket of stocks making up the S&P 500 index. An ETF, however, is not a mutual fund. It trades like stocks and its price changes throughout the day, reflecting its own supply and demand in the market and having its own trading volume. Over the past 20 years, the ETF market has gained tremendous growth. According to ETFGI LLC, the global ETF industry had 3577 ETFs with assets of over $2.2 trillion at the end of November 2013. In the most recent decade, the asset under management by ETFs has achieved an annual expansion rate over 30% on average. Besides the growth in size and volume, the ETF market has also become more diversified in terms of the asset classes of their holdings. Currently, it offers exposure not only to the equity market but also to the bond, commodity and currency markets. As a result, although an

[2]An even earlier exchange-traded portfolio basket is the Toronto stock Exchange Index Participations, which first traded in 1989 in Canada.

ETF is an equity, its price dynamics and return-generating mechanism may be influenced by multiple markets depending on which asset class its holdings belong to. Can the random walk model explain the return-generating process of ETFs? Do bond, commodity and currency ETFs behave like stocks or the assets on which they are based? Are there cases where an ETF's price dynamics are different from the index it tracks? We attempt to answer these questions as part of this first chapter.

While documenting and analyzing the various test statistics against the random walk hypothesis for the ETF market such as variance ratios, Ljung-Box Q-stats, Dickey-Fuller t-ratio, etc., we find that the serial correlation of asset return series is the key statistical measurement of non-random walk behaviors. For instance, the variance ratio statistic at lag q can be approximated by

$$M(q) = \frac{2(q-1)}{q}\hat{\rho}_1 + \frac{2(q-2)}{q}\hat{\rho}_2 + \ldots + \frac{2}{q}\hat{\rho}_{q-1} + O(1/n)$$

and the Ljung-Box Q-stats at lag p can be written as

$$Q(p) = n(n+2)\sum_{k=1}^{p} \frac{\hat{\rho}_k^2}{n-k},$$

where n is the sample size and $\hat{\rho}_k$ is the sample autocorrelation at lag k. These two statistics are either the linear combinations of sample autocorrelations or their squares at different lags. In particular, when $q = 2$ and $p = 1$, we have the neat representations

$$M(2) \approx \hat{\rho}_1$$

and

$$Q(1) \approx n\hat{\rho}_1^2.$$

Therefore, a material understanding of market inefficiency requires knowledge on serial correlation of asset returns. This is rather important since if we were able to find methods which profit from the non-random-walk behaviors of prices, then serial correlation would be a statistical way of justifying those gains. A natural yet more fundamental question arises: what are the sources of serial correlations?

Towards answering this question, Boudoukh, Richardson and Whitelaw (1994) documented three schools of thought. The first school, the *loyalists*, believes that markets rationally process information and that autocorrelations come from market frictions such as non-synchronous trading [Lo and MacKinlay (1988)], price discreteness, bid-ask spreads [Roll (1984)], or systematic changes in either inventory holdings or the flow of information. The second school of thought, the *revisionists*, believes that markets are efficient, however, even when markets are frictionless, short-term stock returns can be autocorrelated due to time-varying economic risk premiums [Campbell *et al.* (1993)] resulting from variations in risk factors including past market returns, past size returns or interest rate spreads. The third school of thought, the *heretics*, believes that markets are not rational. They assert that profitable trading strategies do exist, even on a risk-adjusted basis, and that factors like overreaction or partial price adjustment [Anderson *et al.* (2012)] are influential on security prices causing serial correlations. There has been evidence in supporting each school, yet the true sources of autocorrelation remain controversial.

From our investigation on ETFs and their underliers, we find that non-random walk behaviors do exist. In particular, ETFs whose returns exhibit significant daily autocorrelations tend to be influenced by liquidity, relative valuation and governmental policies. The following findings support the proposition that for portfolio based investments such as ETFs as well as equity and fixed income indices, lack of liquidity leads to higher autocorrelations in returns.

(1) ETFs based on equity portfolios constructed using small-cap stocks tend to show higher serial correlation in returns than large-cap portfolios. Small caps have relatively less shares outstanding and less analyst coverage. They are often less frequently traded with a larger spread compared to stocks with large capitalizations. Information on small-cap stocks is less available to the majority of investors. Therefore, they are less liquid than large caps.

(2) Real-estate investment trusts (REITs) historically have strong positive autocorrelations. Research [Han (1991)] showed, based on data in the 1970s and 1980s, that new information in REITs was not made available to all investors at the same time and even if it was, many REIT investors reacted rather slowly to the arrival of new information. As the REIT market matures over time and

becomes more liquid, the historical patterns are no longer observed in major REIT indices and ETFs.

(3) Corporate bond indices show statistically significant positive autocorrelation, especially in the high-yield category, whereas indices based on Treasury debts do not. This can be largely associated with the fact that corporate bonds are less frequently traded. For instance, only about 10% of the bonds making up the iBoxx Dollar Liquid High Yield Index trade on a daily basis.

Relative valuation and governmental policies can affect return autocorrelations in ETFs, as well. The following examples illustrate this observation.

(1) Equity ETFs based on value stocks characterized by low price-to-book ratios, which are often considered as being undervalued relative to their fundamentals, tend to have low levels of autocorrelation. In particular, for the period 2006-2013, they show statistically significant variance ratios that are less than one or negative autocorrelation.

(2) Since the 2007-2008 subprime mortgage financial crisis, the short-term interest rate returns or short-term Treasury bill ETFs returns have exhibited strong mean reversion behavior under the various forms of quantitative easing, which kept the rates low; whereas for the earlier decade 1996-2005 without such government intervention, the short-term interest rate returns show higher levels of autocorrelation which are mostly in the positive territory.

(3) Natural Gas ETF, UNG, showed significant negative autocorrelation for the period 2008-2013, during which the natural gas experienced oversupply and undervaluation. According to the United States Energy Information Administration, the difference between the dollars per million BTU of crude oil and those of natural gas has increased from 0.1 in 1998 to 12.91 in 2011. In addition, we find that low return autocorrelations in UNG are often associated with natural gas futures in strong contango.

(4) Managed currencies such as Chinese Renminbi and Indian Rupee are subject to currency controls. These operations tend to prevent those currencies from appreciating relative to reserved currencies, therefore, often resulting in undervaluation. For the period 2008-2013, ETFs CNY and INR based on those two currencies respectively both exhibited significant negative autocorrelation compared to liberalized currencies in the developed world.

Our results here add to the literature on the efficient market hypothesis; they help us better understand the levels of market maturity, the switching of economic regimes and the impacts of monetary or foreign exchange policies. The organization of the rest of this chapter is as follows. We first introduce the testing methodologies and statistics in Section 1.2 and then apply them to a collection of popular ETFs. The empirical results are summarized in Section 1.3, where the findings on equity, bond, commodity and currency ETFs are discussed in separate subsections. We conclude the chapter in Section 1.4.

1.2 Test statistics

To establish the basic notations, we denote the price of an asset at time k by P_k and its log-price by X_k, i.e. $X_k = \ln P_k$. Then, the maintained hypothesis is that X_k satisfies

$$X_k = \mu + X_{k-1} + w_k, \tag{1.2}$$

where μ is an arbitrary drift parameter and w_k is the random disturbance at time k, which follows a white noise process. One immediate consequence is that, if we compute the variance, σ_m^2, of the log-return of P_k with a lag of m, then the variances normalized by m will all be the same, i.e.

$$\frac{\sigma_m^2}{\sigma_1^2} = m. \tag{1.3}$$

A random walk can also be considered as a discretized version of a diffusion process, which is defined by x_t, an almost surely continuous process in t such that

$$y_t(\omega) = x_t(\omega) - x_0(\omega) - \int_0^t b_s(\omega)ds \tag{1.4}$$

and

$$z_t(\omega) = y_t(\omega)^2 - \int_0^t a_s(\omega)ds \tag{1.5}$$

are both martingales, where a and b are both bounded progressively measurable functions. Equation (1.3) is therefore associated with the direct implication of z_t being a martingale, that is

$$E\ y_t^2 = E\int_0^t a_s(\omega)ds. \tag{1.6}$$

If we further assume $a_s(\omega)$ to be a constant denoted by σ_a^2, that is

$$a_s(\omega) = \sigma_a^2, \tag{1.7}$$

then, we have

$$E\ y_t^2 = t\sigma_a^2. \tag{1.8}$$

In fact, even when $a_t(\omega)$ is stochastic and time-varying, similar relation may still hold. For instance, suppose we set $a_t(\omega)$ to be a continuous GARCH(1,1) type of stochastic process of the form,

$$da_t = \theta(\gamma - a_t)dt + \rho a_t dW_t, \tag{1.9}$$

where W_t is an independent Brownian motion, then it is easy to see that

$$Ea_t = a_0 e^{-\theta t} + \theta\gamma - \theta\gamma e^{-\theta t} \tag{1.10}$$

and

$$E\ y_t^2 = ta_0 + O(t^2). \tag{1.11}$$

Therefore, if t is small and a_0 is close to its long run average, this relationship reduces to approximately Eq. (1.8). As a result, the corresponding discrete version in Eq. (1.3) can be generalized to a much wider class of random walks whose innovations are heteroscedastic and non-normal[3].

To examine whether asset prices are efficient, we will adopt three statistics and test against the random walk hypothesis, including

[3]Lo and MacKinlay (1988) offered such a generalization of the corresponding discrete formulation.

- the Ljung-Box Q-statistic,
- the Dickey-Fuller t-ratio statistic,
- the variance ratio statistic.

These three methods take advantage of slightly different aspects of the time series under investigation. The Ljung-Box test checks whether the innovations w_k's are uncorrelated up to a predetermined number of lags simultaneously. The Dickey-Fuller test evaluates whether X_k process has a unit root or is weakly stationary. The variance ratio test verifies whether variances using different lagged log-returns are the same after normalization. To compare the power of these statistics, Dickey and Fuller (1979), Lo and Mackinlay (1988) separately perform Monte Carlo simulation based tests. They find that the Ljung-Box test is the least powerful and the variance ratio test is the most powerful. However, we consider all three specifications based on the understanding that they use distinctive methodologies to derive test statistics and their asymptotic distributions. In the following subsections, we describe in details four hypothesis tests using the three statistics we have just introduced to understand the efficiency of the ETF market.

1.2.1 *The Ljung-Box test*

The Ljung-Box test [Ljung and Box (1978)] is an improved version of the Portmanteau test proposed earlier by Box and Pierce in 1970. The refinement offers better finite-sample properties. If we denote the sample k-lag autocorrelation as $\hat{\rho}_k$, then the Ljung-Box Q-statistic is defined as

$$Q(p) = n(n+2)\sum_{k=1}^{p} \frac{\hat{\rho}_k^2}{n-k}, \tag{1.12}$$

and it is associated with the null hypothesis that

$$\rho_1 = \rho_2 = \ldots = \rho_p = 0, \tag{1.13}$$

where n is the sample size and p is the number of lags tested.

$Q(p)$ is asymptotically distributed as a chi-squared distribution with p degrees of freedom. For a significance level α, the critical region for rejection of the null hypothesis is

$$Q(p) > \chi^2_{1-\alpha,p}, \tag{1.14}$$

where $\chi^2_{1-\alpha,p}$ is the α-quantile of the chi-squared distribution with p degrees of freedom. For our purpose, the autocorrelation function is computed on $X_k - X_{k-1} - \hat{\mu}$, with $\hat{\mu}$ being the sample mean log-return of the price series. Rejecting the null hypothesis for this test will suggest significant autocorrelations in asset returns, disproving the random walk model.

1.2.2 *The Dickey-Fuller test*

The Dickey-Fuller test (1979) is also called the unit-root test which evaluates whether the log-price X_t follows a random walk or is stationary. There are in general two specifications, with or without drift. In our context, we use the former, that is

$$X_t = \mu + \phi X_{t-1} + w_t, \tag{1.15}$$

where μ is an arbitrary drift parameter and w_t is the random disturbance at time t, which follows a white noise process. The null hypothesis is $\phi = 1$ and the alternative hypothesis is $\phi < 1$. The test statistic is defined by the t-ratio,

$$t\text{-ratio} = \frac{\hat{\phi} - 1}{\hat{\sigma}_\phi}, \tag{1.16}$$

where $\hat{\phi}$ and $\hat{\sigma}_\phi$ are the least square estimators. Dickey and Fuller show that when $\mu \neq 0$ the t-ratio is asymptotically normal distributed. Rejection of this null hypothesis will indicate that the log price time series does not contain a unit root and may very likely be weakly stationary.

1.2.3 *The variance ratio test*

Lo and MacKinlay (1988) test the random walk hypothesis for weekly stock market returns by comparing variance estimators derived from data sampled at different lags for the period 1962-1985. They propose a variance ratio statistic that is sensitive to serial correlation but robust to heteroscedastic innovations and nonnormality. Based on this statistic, the random walk model is strongly rejected for both individual and portfolio of CRSP stocks over the entire sample period and subperiods.

Suppose we are considering $nq+1$ data points denoted by $X_0, X_1, ..., X_{nq}$ consistently as before. Under the assumption that w_t is an independent and identically distributed normal random variable, i.e. $w_t \sim i.i.d.\ N(0, \sigma_0^2)$, the mean estimator is

$$\hat{\mu} = \frac{1}{nq}\sum_{k=1}^{nq}(X_k - X_{k-1}) = \frac{1}{nq}(X_{nq} - X_0), \tag{1.17}$$

the unbiased variance estimator for the one-period log-return is given by,

$$\hat{\sigma}_a^2 = \frac{1}{nq-1}\sum_{k=1}^{nq}(X_k - X_{k-1} - \hat{\mu})^2, \tag{1.18}$$

and the unbiased variance estimator for the q-period log-return is given by,

$$\hat{\sigma}_c^2(q) = \frac{1}{m}\sum_{k=q}^{nq}(X_k - X_{k-q} - q\hat{\mu})^2, \tag{1.19}$$

where $m = q(nq - q + 1)\left(1 - \frac{1}{n}\right)$. Define

$$M(q) = \frac{\hat{\sigma}_c^2(q)}{\hat{\sigma}_a^2} - 1. \tag{1.20}$$

Lo and MacKinlay show that the statistic, $M(q)$, is asymptotically normal, that is

$$\sqrt{\frac{3nq^2}{2(2q-1)(q-1)}}M(q) \sim N(0,1). \tag{1.21}$$

In addition, they derive a version of the test statistic that is rather general, in that it is robust to heteroscedastic but uncorrelated innovations. Under mild assumptions, it can be shown that the following equation holds asymptotically

$$M(q) = \sum_{j=1}^{q-1}\frac{2(q-j)}{q}\hat{\rho}(j). \tag{1.22}$$

Moreover, if we denote by $\delta(j)$ the asymptotic variances of the j-th order autocorrelation, $\hat{\rho}(j)$, for each j and $\theta(q)$ the asymptotic variances of $M(q)$. Then $M(q)$ converges to 0 almost surely as $n \to \infty$ and

$$z(q) = \sqrt{\frac{1}{\hat{\theta}(q)}}M(q) \sim N(0,1), \tag{1.23}$$

where $\hat{\theta}(q)$ is a heteroscedasticity-consistent estimator of $\theta(q)$ defined by

$$\hat{\theta}(q) = \sum_{j=1}^{q-1} \left[\frac{2(q-j)}{q} \right]^2 \hat{\delta}(j) \tag{1.24}$$

and $\hat{\delta}(j)$ is a heteroscedasticity-consistent estimator of $\delta(j)$ defined by

$$\hat{\delta}(j) = \frac{\sum_{k=j+1}^{nq} (X_k - X_{k-1} - \hat{\mu})^2 (X_{k-j} - X_{k-j-1} - \hat{\mu})^2}{[\sum_{k=1}^{nq} (X_k - X_{k-1} - \hat{\mu})^2]^2}. \tag{1.25}$$

Note, for the case $q = 2$, we may expand the formula of $M(2)$ and write explicitly

$$M(2) = \frac{\hat{\sigma_c}^2(2)}{\hat{\sigma_a}^2} - 1 \tag{1.26}$$

$$= \frac{n}{n-1} \frac{\sum_{k=2}^{2n} (X_k - X_{k-1} - \hat{\mu})(X_{k-1} - X_{k-2} - \hat{\mu})}{\hat{\sigma_a}^2}, \tag{1.27}$$

which is the sample autocorrelation function at lag 1. Since sample variances computed using returns at different frequencies will be the same after normalization if the random walk assumption is true, the rejection of the null hypothesis for this test will disprove the random walk model as well.

1.2.4 *Ljung-Box test on GARCH residuals*

Statistically speaking, when any of the above three tests succeeds in rejecting its own null hypothesis, we have reasons to believe that the log-price process deviates from the random walk hypothesis. Among the many combinations of the outcomes of these tests, one particular situation may occur, where the Dickey-Fuller test fails to reject the unit root hypothesis but the Ljung-Box test succeeds in rejecting the independence of the log-return series. When this happens, we may further fit the return series to a GARCH model with innovations following the Student-t distribution so as to better understand the price-generating process. After all, testing market efficiency is model-specific. As shown by LeRoy (1973), equilibrium prices under rational expectations do not necessarily form martingales. We believe that one major reason behind the rejection of uncorrelated innovations is the volatility clustering in log-returns. Therefore, we set

$$X_k - X_{k-1} - \hat{\mu} = \sqrt{v_k} e_k \tag{1.28}$$

where e_k's are independent, identically distributed Student-t noises with d degree of freedom and v_t is the conditional variance satisfying a GARCH(1,1) model specified by

$$v_{k+1} = \alpha + \beta v_k e_k^2 + \gamma v_k. \tag{1.29}$$

To determine the parameters α, β and γ, we may maximize the log-likelihood function $l_n(\alpha, \beta, \gamma, d)$ defined by

$$\begin{aligned} l_n(\alpha, \beta, \gamma, d) = \sum_{k=1}^{n} \ln \Gamma\left(\frac{d+1}{2}\right) - \ln \Gamma\left(\frac{d}{2}\right) - \frac{1}{2}\ln(d\pi - 2\pi) \\ - \frac{1}{2}\ln v_k - \left(\frac{d+1}{2}\right)\ln\left(1 + \frac{(X_k - X_{k-1})^2}{v_k(d-2)}\right), \end{aligned} \tag{1.30}$$

where n is the number of observations available. Once the parameters are fitted, we apply the Ljung-Box test on e_t for independence to validate this price-generating process hypothesized jointly by

$$\begin{aligned} P_k &= P_{k-1}\exp\left(\hat{\mu} + \sqrt{v_k}e_k\right) \\ v_{k+1} &= \alpha + \beta v_k e_k^2 + \gamma v_k, \end{aligned} \tag{1.31}$$

where e_k are independent and identically distributed innovations and $\hat{\mu}$ is the sample variance of log-returns.

1.3 Summary of findings

In this section, we summarize our findings on the seraial correlation in ETF returns. The test results using the three statistics described in the previous section on representative ETFs over the period Jan 1, 2006 - Sep 30, 2013 are documented and analyzed[4]. The related underlying assets and indices are investigated as well when they offer insights into explaining the levels of autocorrelation. Equity, bond, commodity and currency ETFs are discussed in separate subsections. The goal is to understand where evidences against the random walk hypothesis are mostly concentrated within each group and how we can justify them. We do not consider inverse and leveraged ETFs because they are somewhat different in the way they are constructed.

[4]We choose to start the sample period in the year of 2006 because the inception dates of most of the stocks and ETFs we consider, especially equity ETFs, fall ahead of this year but not earlier years. Therefore, Jan 1, 2006 seems to be a sensible cutoff. Still, many fixed income and currency based ETFs were created later in 2007 and 2008. For those ETFs, we use the period Jan 1, 2008 - Sep 30, 2013.

1.3.1 *Equity ETFs*

Equity ETFs are the oldest ETFs. Compared to other similar pooled investments such as equity mutual funds, these ETFs tend to be inexpensive, with low expense ratios, tax-saving and fairly narrowed bid-ask spreads. As a result, equity ETFs have achieved rather rapid growth. Figure 1.1 plots the total assets under management by the equity ETFs available in the US market for the period 2005-2013. We see that in nine years, the asset under management (AUM) of equity ETFs has grown from $200 billion to over $1.3 trillion in the US.

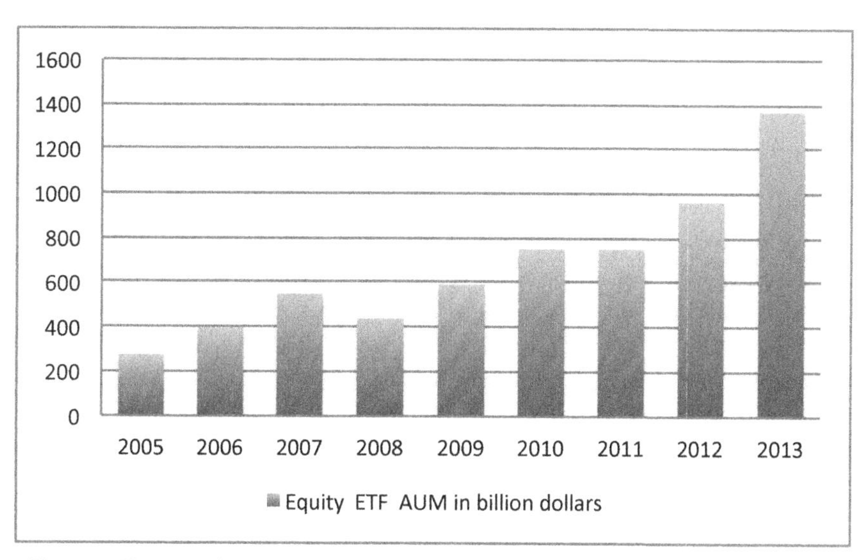

Fig. 1.1: The total AUMs of the US equity ETFs by year retrieved from Bloomberg.

Equity ETFs were first created to offer easier ways to trade diversified portfolios based on only the US broad market indices. For instance, SPY (SPDR S&P 500 Index ETF) offers exposure to about 500 major domestic stocks, covering about 75% of the US market by market capitalization. IWV (iShares Russell 3000 Index Fund) allows investors to trade over 3000 stocks in a single fund in the US market. As the ETF market expands, even broader market based ETFs were subsequently created. For example, apart from the US market, now an investor can achieve global diversification by investing in the all world ETFs such as ACWI (iShares MSCI All Country

World Index Fund) and VT (Vanguard Total World Stock ETF) which cover both domestic and international equities. Furthermore, it is often perceived that stocks in the developed markets often have considerably different characteristics than stocks in the emerging markets. ETFs also allow investors to trade for instance the emerging markets as a whole. An example in this category is EEM (iShares MSCI Emerging Markets Index Fund), which has turned into one of the most liquid exchange traded funds in the US equity market.

As the ETF market receives more and more attention and popularity, ETF managers have also started creating funds which target specific well-established market factors, making them easily tradable like any single stocks out there on the exchange. Many factors now have their very own realized price series instead of being derived as certain weighted averages of the prices of their constituents.

Size and style[5] have been established as two essential factors in asset pricing by Fama and French (1992). They collect NYSE, AMEX and NASDAQ stocks and allocate them to 10×10 portfolios ranked by market capitalization and book-to-market ratios for the period July 1963 - December 1990. Their empirical results show that the average monthly cross-section returns increase in the book-to-market ratio and decrease in market-capitalization. In addition, they create Fama-French three-factor model [Fama and French (1993)] which extends the classic CAPM model by adding SMB (small-cap minus big-cap) and HML (high BM ratio minus low BM ratio) factors. The new model explains over 90% of the diversified portfolio returns, whereas CAPM explains on average 70%. Higher expected returns from small-cap value stocks are arguably explained by their having higher expected risks, though later literature disagrees with the risk premium argument [Porta *et al.* (1997)][6].

The notable findings of Fama and French naturally give incentives to the ETF sponsors to create index ETFs which track equity portfolios con-

[5]Fama and French primarily characterize the style factor by the book-to-market ratios. They also considered earnings-price ratio and leverage ratios.

[6]Lakonishok et al. showed that the outperformance of value stocks over growth stocks is largely attributable to the fact that the earning surprises for values stocks tend to be systematically more positive than those of growth stocks. This is in disagreement with the risk premium argument.

structed by size and style. We test if size and style remain factors for the autocorrelation of these ETFs' daily returns. The following of this section shows this is indeed true.

1.3.1.1 *Classification by size*

We first consider 9 popular broad market equity index ETFs by three major fund managers Vanguard, iShares and SPDR. These 9 ETFs are characterized by the market capitalizations or sizes[7] of the stocks making up the funds' portfolios. We find that the rejection of the random walk hypothesis is the strongest among the large-cap and small-cap ETFs.

Table 1.1 summarizes the test statistics of 9 ETFs which track major US equity indices categorized by sizes for the period Jan 1, 2006 - Sep 30, 2013. The variance ratios $1 + M(q)$, $q = 2, 5, 10, 20$, are reported in VR labeled columns and in the main rows in line with the ETF tickers, whereas the corresponding z-scores are shown in parentheses immediately below each main row. The DF labeled column shows the first lag regression coefficients of log prices for the Dickey-Fuller test in the main rows with the corresponding p-values in parentheses immediately below the main rows. The LB labeled columns represent the Q-statistics for the Ljung-Box test on the innovations w_k in the main rows with p-values in parentheses immediately below each main row. Similarly, the GLB labeled columns show the Q-statistics for the Ljung-Box test on the standardized residual e_k from fitting returns to a GARCH(1,1) model. Significance levels at 99% and 95% are signaled by two and one stars next to the main row statistics respectively. Under the random walk null hypothesis, the value of the variance ratio is 1, the Dickey-Fuller regression first lag coefficient is 1 and w_t's are independent. Under the random walk with GARCH residuals price hypothesis, the Dickey-Fuller regression first lag coefficient is 1, LB Q-statistics are significant and GLB Q-statistics are insignificant.

From the table, we see that the Ljung-Box test and the Dickey-Fuller

[7]There is no universal consensus on the exact definitions of the various market caps, but typically we have the following practices. Giant Cap : more than \$200 billion, Large Cap : \$10 billion $\sim$ \$200 billion, Mid Cap : \$2 billion $\sim$ \$10 billion, Small Cap : \$300 million $\sim$ \$2 billion, Micro Cap : less than \$300 million.

Category	Ticker	VR(2)	VR(5)	VR(10)	VR(20)	DF	LB(5)	LB(10)	GLB(5)	GLB(10)
Large	VV	0.91*	0.81*	0.73	0.73	1.00	34.15**	43.25**	10.18	17.88
		(-2.28)	(-1.98)	(-1.75)	(-1.20)	(0.72)	(0.00)	(0.00)	(0.07)	(0.06)
	SPY	0.91*	0.78*	0.71	0.69	1.00	40.15**	45.51**	10.90	19.06*
		(-2.15)	(-2.02)	(-1.76)	(-1.25)	(0.69)	(0.00)	(0.00)	(0.05)	(0.04)
	IWB	0.91*	0.82	0.75	0.75	1.00	30.94**	38.42**	10.54	18.18
		(-2.15)	(-1.83)	(-1.64)	(-1.11)	(0.73)	(0.00)	(0.00)	(0.06)	(0.05)
Mid	VO	0.95	0.87	0.80	0.82	1.00	14.43*	23.40**	4.68	11.15
		(-1.29)	(-1.41)	(-1.38)	(-0.84)	(0.75)	(0.01)	(0.01)	(0.46)	(0.35)
	MDY	0.94	0.84	0.75	0.76	1.00	27.35**	38.05**	6.23	11.42
		(-1.46)	(-1.57)	(-1.55)	(-1.02)	(0.72)	(0.00)	(0.00)	(0.28)	(0.33)
	IWR	0.96	0.90	0.84	0.88	1.00	11.78*	18.07	5.88	12.83
		(-1.00)	(-1.14)	(-1.15)	(-0.59)	(0.79)	(0.04)	(0.05)	(0.32)	(0.23)
Small	VB	0.94	0.88	0.80	0.82	1.00	18.76**	30.28**	5.59	10.64
		(-1.49)	(-1.39)	(-1.46)	(-0.91)	(0.73)	(0.00)	(0.00)	(0.35)	(0.39)
	SLY	0.91	0.75	0.67	0.68	1.00	29.85**	39.03**	9.60	10.63
		(-0.78)	(-1.23)	(-1.32)	(-1.07)	(0.67)	(0.00)	(0.00)	(0.09)	(0.39)
	IWM	0.91*	0.84	0.75	0.75	1.00	25.03**	34.26**	8.05	12.04
		(-2.21)	(-1.87)	(-1.82)	(-1.27)	(0.61)	(0.00)	(0.00)	(0.15)	(0.28)

Table 1.1: The test statistics of 9 ETFs which track major US equity indices categorized by sizes for the period Jan 1, 2006 - Sep 30, 2013. The variance ratios $1 + M(q)$, $q = 2, 5, 10, 20$, are reported in VR labeled columns and in the main rows in line with the ETF tickers, whereas the corresponding z-scores are shown in parentheses immediately below each main row. The DF labeled column shows the first lag regression coefficients of log prices for the Dickey-Fuller test in the main rows with the corresponding p-values in parentheses immediately below the main rows. The LB labeled columns represent the Q-statistics for the Ljung-Box test on the innovations w_k in the main rows with p-values in parentheses immediately below each main row. Similarly, the GLB labeled columns show the Q-statistics for the Ljung-Box test on the standardized residual e_k from fitting returns to a GARCH(1,1) model. Significance levels at 99% and 95% are signaled by two and one stars next to the main row statistics respectively. Under the random walk null hypothesis, the value of the variance ratio is 1, the Dickey-Fuller regression first lag coefficient is 1 and w_t's are independent. Under the random walk with GARCH residuals price hypothesis, the Dickey-Fuller regression first lag coefficient is 1, LB Q-statistics are significant and GLB Q-statistics are insignificant.

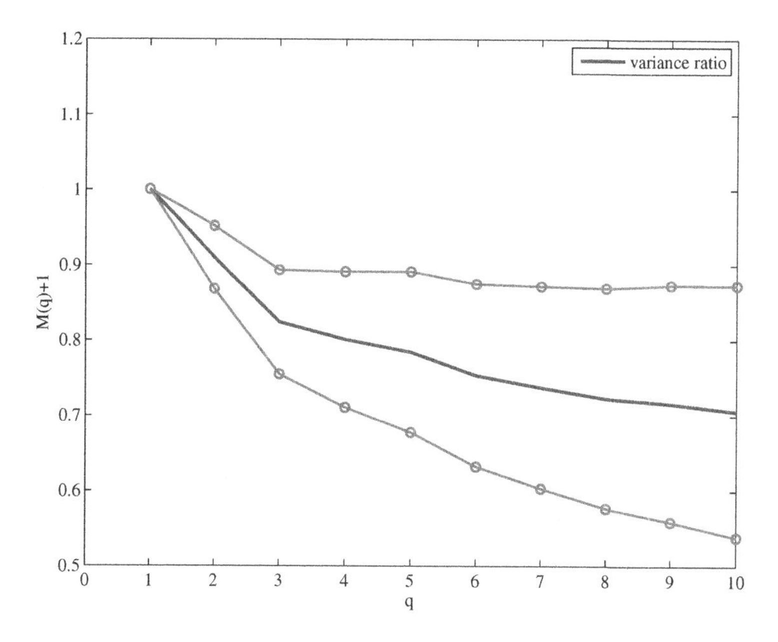

Fig. 1.2: The variance ratio $M(q) + 1$ versus q for SPY as the solid line. The solid lines marked with circles are the one standard deviation confidence band.

test do not seem to help distinguish the 9 ETFs by size. The Ljung-Box statistics in columns labeled by LB(5) and LB(10) indicate the existence of autocorrelation in the log-returns of all 9 ETFs with IWR being the only case where LB(10) is not statistically significant. However, after filtering out the heteroscedasticity by fitting the conditional variance of log-returns to the GARCH(1,1) model with Student-t distributed innovations, the information contained in the price series is mostly removed; as seen from the GLB labeled columns, all but one of the p-values are greater than 0.05. The Dickey-Fuller statistics are all insignificant, suggesting little evidence that the first lag coefficients in the regression of $\ln P_k$ on $\ln P_{k-1}$ are different from one. That is, we do not have statistical evidence in rejecting the non-stationarity of the ETF's log-price.

The variance ratio statistics we computed in the table are robust under heteroscedasticity and non-normality in log-returns and they tell the ETFs apart by size. The test statistics $M(q)+1$ and their z-scores are summarized in the columns labeled by $VR(q)$, where q is the lag of the log-returns used.

We see that $M(q)+1$ is typically a decreasing function in q. The z-score's absolute levels also decline in q. All 3 large-cap ETFs have significant variance ratios for $q=2$ or 5, none of the mid-cap ETFs have significant variance ratios and only one small-cap ETF's variance ratio is significant at lag 2. Figure 1.2 plots lag numbers versus $M(q)+1$ (the solid line), with confidence bands for the ETF SPY (the solid lines marked with circles). This figure also represents the common behaviors we see in Table 1.1 of other ETFs.

1.3.1.2 *Classification by size and style*

Table 1.2 offers a more refined analysis and uses 18 ETFs which track major US equity indices categorized by both size and style into 6 groups. The style factor divides companies into two kinds: value or growth. A value stock tends to trade at a lower price relative to its fundamentals such as earnings, sales, etc. As a result, investors often consider value stocks undervalued. Value stocks usually have high dividend yields and low price-to-book ratios. Growth stocks typically do not pay dividends and reinvest all of their retained earnings in capital projects. They often own new or breakthrough products, which lead to faster growth rates than average companies within the same industry. Many technology companies are of the growth style.

The test statistics overall suggest that for this period value stocks and large-cap stocks tend to behave less like a random walk. From Table 1.2, we observe that all 3 large-cap value ETFs have significant statistics from most tests we implemented, exhibiting the strongest evidence against the random walk hypothesis. In terms of variance ratios, all growth stocks based ETFs failed to reject the null hypothesis except one with a large market capitalization named IVW; none mid-cap stocks based ETFs possess statistically significant variance ratios. We also observe that small-cap value ETFs give strong evidence rejecting the random walk hypothesis; however, they differ from the large-cap value ETFs in the GLB columns, where the Ljung-Box Q-statistics on the residuals from fitting the GARCH(1,1) model to log-return's conditional variance is shown. The significant GLB Q-statistics may suggest that the price-generating process of large-cap value ETFs is the most complex among the 6 categories listed in this table. In addition,

Category	Ticker	VR(2)	VR(5)	VR(10)	VR(20)	DF	LB(5)	LB(10)	GLB(5)	GLB(10)
Large	VTV	0.89**	0.79*	0.71	0.69	1.00	35.54**	48.66**	11.53*	22.63*
Value		(-2.59)	(-2.22)	(-1.95)	(-1.38)	(0.60)	(0.00)	(0.00)	(0.04)	(0.01)
	IVE	0.91*	0.83	0.75	0.74	1.00	28.46**	39.76**	12.94*	23.28**
		(-2.21)	(-1.85)	(-1.73)	(-1.20)	(0.61)	(0.00)	(0.00)	(0.02)	(0.01)
	IWD	0.90*	0.81	0.72	0.71	1.00	35.71**	45.76**	11.36*	19.38*
		(-2.47)	(-1.94)	(-1.80)	(-1.28)	(0.61)	(0.00)	(0.00)	(0.04)	(0.04)
Large	VUG	0.92	0.83	0.76	0.78	1.00	30.19**	37.37**	6.43	11.62
Growth		(-1.81)	(-1.66)	(-1.50)	(-0.97)	(0.82)	(0.00)	(0.00)	(0.27)	(0.31)
	IVW	0.90*	0.79*	0.72	0.72	1.00	36.87**	45.34**	11.18*	17.68
		(-2.41)	(-2.11)	(-1.79)	(-1.20)	(0.81)	(0.00)	(0.00)	(0.05)	(0.06)
	IWF	0.93	0.84	0.77	0.78	1.00	27.23**	32.39**	6.79	12.29
		(-1.70)	(-1.64)	(-1.47)	(-0.95)	(0.82)	(0.00)	(0.00)	(0.24)	(0.27)
Mid	VOE	0.93	0.85	0.78	0.82	1.00	15.55**	24.03**	5.98	13.68
Value		(-1.68)	(-1.58)	(-1.46)	(-0.81)	(0.76)	(0.01)	(0.01)	(0.31)	(0.19)
	IJJ	0.93	0.84	0.77	0.79	1.00	21.40**	32.72**	5.67	11.49
		(-1.79)	(-1.70)	(-1.57)	(-0.96)	(0.68)	(0.00)	(0.00)	(0.34)	(0.32)
	IWS	0.94	0.87	0.82	0.86	1.00	15.28**	28.84**	7.06	17.47
		(-1.57)	(-1.39)	(-1.29)	(-0.67)	(0.74)	(0.01)	(0.00)	(0.22)	(0.06)
Mid	VOT	0.97	0.89	0.81	0.83	1.00	13.54*	19.49*	6.30	10.84
Growth		(-0.63)	(-1.11)	(-1.19)	(-0.73)	(0.74)	(0.02)	(0.03)	(0.28)	(0.37)
	IJK	0.98	0.90	0.82	0.84	1.00	14.26*	22.63*	6.27	10.09
		(-0.60)	(-1.09)	(-1.27)	(-0.79)	(0.82)	(0.01)	(0.01)	(0.28)	(0.43)
	IWP	0.97	0.88	0.82	0.85	1.00	10.83	17.32	4.58	11.05
		(-0.95)	(-1.29)	(-1.22)	(-0.68)	(0.79)	(0.05)	(0.07)	(0.47)	(0.35)
Small	VBR	0.91*	0.84	0.76	0.79	1.00	26.86**	38.50**	6.99	12.64
Value		(-2.27)	(-1.80)	(-1.69)	(-1.02)	(0.62)	(0.00)	(0.00)	(0.22)	(0.24)
	IJS	0.90*	0.81*	0.73*	0.73	1.00	30.23**	38.15**	10.79	14.96
		(-2.55)	(-2.18)	(-2.05)	(-1.39)	(0.61)	(0.00)	(0.00)	(0.06)	(0.13)
	IWN	0.89**	0.80*	0.72*	0.72	1.00	34.62**	48.61**	10.33	15.79
		(-2.77)	(-2.28)	(-2.07)	(-1.40)	(0.47)	(0.00)	(0.00)	(0.07)	(0.11)
Small	VBK	0.97	0.92	0.84	0.86	1.00	12.25*	20.17*	4.27	8.44
Growth		(-0.81)	(-0.94)	(-1.15)	(-0.71)	(0.81)	(0.03)	(0.03)	(0.51)	(0.59)
	IJT	0.94	0.87	0.79	0.80	1.00	15.87**	24.45**	5.90	11.31
		(-1.54)	(-1.51)	(-1.55)	(-0.99)	(0.83)	(0.01)	(0.01)	(0.32)	(0.33)
	IWO	0.94	0.88	0.80	0.79	1.00	19.36**	27.39**	5.38	9.41
		(-1.64)	(-1.35)	(-1.54)	(-1.10)	(0.73)	(0.00)	(0.00)	(0.37)	(0.49)

Table 1.2: The test statistics of 18 ETFs which track major US equity indices categorized by sizes and styles for the period Jan 1, 2006 - Sep 30, 2013. The variance ratios $1 + M(q)$, $q = 2, 5, 10, 20$, are reported in VR labeled columns and in the main rows in line with the ETF tickers, whereas the corresponding z-scores are shown in parentheses immediately below each main row. The LB labeled columns represent the Q-statistics for the Ljung-Box test on the innovations w_k in the main rows with p-values in parentheses immediately below each main row. Similarly, the GLB labeled columns show the Q-statistics for the Ljung-Box test on the standardized residual e_k from fitting returns to a GARCH(1,1) model. Significance levels at 99% and 95% are signaled by two and one stars next to the main row statistics respectively. Under the random walk null hypothesis, the value of the variance ratio is 1, the Dickey-Fuller regression first lag coefficient is 1 and w_t's are independent. Under the random walk with GARCH residuals price hypothesis, the Dickey-Fuller regression first lag coefficient is 1, LB Q-statistics are significant and GLB Q-statistics are insignificant.

by comparing the variance ratios between the small-cap value ETFs with the small-cap growth ETFs, we see that all rejections of the null hypothesis occur on the side of the value ETFs. This implies that the small stocks effect we observed in Table 1.1 may primarily derive from the value portion of the ETF holdings. The Dickey-Fuller test failed again to establish stationarity in the log-price series over all the 18 ETFs considered here. The monotonicity of the variance ratio statistics in q remains valid. All variance ratios are less than 1 and rejections of null hypothesis take place mostly with $q = 2$ and 5 if they occurred.

1.3.1.3 *Classification by sector*

The empirical findings we have obtained so far in Table 1.2 indicates that size and style may be important factors for autocorrelations or more generally variance ratios, as well. We further investigate in this direction by testing the efficiency on sector ETFs. We first study how sector ETFs' log-price series fit the random walk hypothesis and how idiosyncratic sector characteristics may affect their test statistics. Then, we break down sector ETFs by size and style factors and research on factor impacts.

Tables 1.3, 1.4 and 1.5 report the test statistics of totally 32 sector ETFs covering 11 major US equity sectors. If we classify them by strong rejection (I), mild rejection (II) and no rejection (III) three categories, then the sector ETFs can be roughly listed as follows,

I : consumer staples, financials
II : energy, healthcare, technology, telecom, REIT[8], utilities
III : consumer discretionary, industrial, material.

Group I ETFs have null hypothesis rejections in all except for the Dickey-Fuller tests. Their price-generating processes seem to be the most complicated among all sector ETFs. The serial correlations in their log-returns

[8]In the SPDR series sector ETFs, REIT stocks are classified as part of Financials in XLF. REIT is short for Real Estate Investment Trust. A REIT is a security which sells like a stock on the major exchanges and invests in real estate through properties or mortgages. It usually receives special tax considerations and offers investors high yields and highly liquid method of investing in real estate.

Category	Ticker	VR(2)	VR(5)	VR(10)	VR(20)	DF	LB(5)	LB(10)	GLB(5)	GLB(10)
Consumer	IYC	0.96	0.89	0.85	0.88	1.00	13.52*	19.66*	5.48	12.95
Discretionary		(-0.98)	(-1.26)	(-1.13)	(-0.62)	(0.94)	(0.02)	(0.03)	(0.36)	(0.23)
	XLY	0.96	0.90	0.84	0.86	1.00	8.94	14.60	4.45	14.94
		(-0.93)	(-1.15)	(-1.12)	(-0.68)	(0.92)	(0.11)	(0.15)	(0.49)	(0.13)
	VCR	1.01	0.95	0.92	0.96	1.00	8.73	12.88	5.56	12.11
		(0.18)	(-0.53)	(-0.60)	(-0.20)	(0.93)	(0.12)	(0.23)	(0.35)	(0.28)
Consumer	IYK	0.92	0.79*	0.72	0.75	1.00	45.80**	52.48**	11.41*	17.82
Staples		(-1.73)	(-2.03)	(-1.77)	(-1.11)	(0.92)	(0.00)	(0.00)	(0.04)	(0.06)
	XLP	0.90**	0.79*	0.68*	0.69	1.00	41.91**	47.83**	18.08**	25.90**
		(-2.60)	(-2.43)	(-2.43)	(-1.66)	(0.92)	(0.00)	(0.00)	(0.00)	(0.00)
	VDC	0.90*	0.77*	0.69	0.70	1.00	53.21**	62.20**	13.23*	18.97*
		(-2.17)	(-2.17)	(-1.92)	(-1.30)	(0.92)	(0.00)	(0.00)	(0.02)	(0.04)
Energy	IYE	0.87	0.75	0.69	0.63	0.99	71.32**	78.68**	4.24	10.10
		(-1.89)	(-1.73)	(-1.51)	(-1.32)	(0.16)	(0.00)	(0.00)	(0.52)	(0.43)
	XLE	0.91*	0.78*	0.71	0.65	0.99	36.90**	46.26**	5.75	11.27
		(-2.18)	(-2.17)	(-1.80)	(-1.49)	(0.20)	(0.00)	(0.00)	(0.33)	(0.34)
	VDE	0.92	0.81	0.74	0.68	0.99	37.54**	46.94**	2.75	7.31
		(-1.83)	(-1.79)	(-1.55)	(-1.32)	(0.19)	(0.00)	(0.00)	(0.74)	(0.70)
Financial	IYF	0.89*	0.80*	0.67*	0.64	1.00	38.60**	52.96**	11.86*	22.49*
		(-2.38)	(-2.04)	(-2.12)	(-1.58)	(0.43)	(0.00)	(0.00)	(0.04)	(0.01)
	XLF	0.88*	0.78*	0.65*	0.62	1.00	40.76**	57.97**	16.13**	25.88**
		(-2.49)	(-2.16)	(-2.20)	(-1.65)	(0.42)	(0.00)	(0.00)	(0.01)	(0.00)
	VFH	0.89*	0.79*	0.68*	0.65	1.00	38.16**	53.25**	11.06	21.23*
		(-2.47)	(-2.09)	(-2.09)	(-1.57)	(0.41)	(0.00)	(0.00)	(0.05)	(0.02)

Table 1.3: The test statistics of 12 ETFs which track major US equity sector indices for the period Jan 1, 2006 - Sep 30, 2013. The variance ratios $1 + M(q)$, $q = 2, 5, 10, 20$, are reported in VR labeled columns and in the main rows in line with the ETF tickers, whereas the corresponding z-scores are shown in parentheses immediately below each main row. The DF labeled column shows the first lag regression coefficients of log prices for the Dickey-Fuller test in the main rows with the corresponding p-values in parentheses immediately below the main rows. The LB labeled columns represent the Q-statistics for the Ljung-Box test on the innovations w_k in the main rows with p-values in parentheses immediately below each main row. Similarly, the GLB labeled columns show the Q-statistics for the Ljung-Box test on the standardized residual e_k from fitting returns to a GARCH(1,1) model. Significance levels at 99% and 95% are signaled by two and one stars next to the main row statistics respectively. Under the random walk null hypothesis, the value of the variance ratio is 1, the Dickey-Fuller regression first lag coefficient is 1 and w_t's are independent.

Category	Ticker	VR(2)	VR(5)	VR(10)	VR(20)	DF	LB(5)	LB(10)	GLB(5)	GLB(10)
Healthcare	IYH	0.94	0.87	0.83	0.81	1.00	29.10**	36.29**	5.32	10.44
		(-1.46)	(-1.28)	(-1.14)	(-0.86)	(0.96)	(0.00)	(0.00)	(0.38)	(0.40)
	XLV	0.93	0.84	0.80	0.78	1.00	33.57**	43.17**	6.90	14.04
		(-1.51)	(-1.25)	(-1.09)	(-0.90)	(0.95)	(0.00)	(0.00)	(0.23)	(0.17)
	VHT	0.94	0.86	0.81	0.79	1.00	34.97**	43.62**	6.00	10.82
		(-1.46)	(-1.30)	(-1.20)	(-0.91)	(0.96)	(0.00)	(0.00)	(0.31)	(0.37)
Industrial	IYJ	0.97	0.93	0.89	0.93	1.00	10.47	21.05*	5.86	15.18
		(-0.76)	(-0.84)	(-0.87)	(-0.36)	(0.75)	(0.06)	(0.02)	(0.32)	(0.13)
	XLI	0.95	0.90	0.87	0.92	1.00	7.13	14.60	5.77	12.14
		(-1.55)	(-1.23)	(-1.01)	(-0.44)	(0.72)	(0.21)	(0.15)	(0.33)	(0.28)
	VIS	0.97	0.93	0.89	0.92	1.00	11.99*	22.90*	5.69	13.55
		(-0.85)	(-0.79)	(-0.86)	(-0.42)	(0.70)	(0.03)	(0.01)	(0.34)	(0.19)
Material	IYM	0.98	0.92	0.86	0.86	1.00	8.24	16.99	4.35	9.44
		(-0.54)	(-0.91)	(-1.00)	(-0.68)	(0.29)	(0.14)	(0.07)	(0.50)	(0.49)
	XLB	0.97	0.88	0.83	0.79	1.00	8.29	11.71	4.42	7.43
		(-1.00)	(-1.42)	(-1.35)	(-1.08)	(0.30)	(0.14)	(0.30)	(0.49)	(0.68)
	VAW	0.97	0.91	0.85	0.84	1.00	9.56	16.76	4.76	8.62
		(-0.71)	(-1.07)	(-1.11)	(-0.78)	(0.40)	(0.09)	(0.08)	(0.45)	(0.57)
REIT	IYR	0.81**	0.66**	0.59*	0.61	1.00	76.80**	93.26**	3.41	8.54
		(-3.53)	(-3.02)	(-2.36)	(-1.51)	(0.36)	(0.00)	(0.00)	(0.64)	(0.58)
	VNQ	0.80**	0.66**	0.59*	0.60	1.00	90.20**	103.91**	4.53	9.94
		(-3.92)	(-3.07)	(-2.37)	(-1.55)	(0.36)	(0.00)	(0.00)	(0.48)	(0.45)

Table 1.4: The test statistics of 11 ETFs which track major US equity sector indices for the period Jan 1, 2006 - Sep 30, 2013. The variance ratios $1 + M(q)$, $q = 2, 5, 10, 20$, are reported in VR labeled columns and in the main rows in line with the ETF tickers, whereas the corresponding z-scores are shown in parentheses immediately below each main row. The DF labeled column shows the first lag regression coefficients of log prices for the Dickey-Fuller test in the main rows with the corresponding p-values in parentheses immediately below the main rows. The LB labeled columns represent the Q-statistics for the Ljung-Box test on the innovations w_k in the main rows with p-values in parentheses immediately below each main row. Similarly, the GLB labeled columns show the Q-statistics for the Ljung-Box test on the standardized residual e_k from fitting returns to a GARCH(1,1) model.

Category	Ticker	VR(2)	VR(5)	VR(10)	VR(20)	DF	LB(5)	LB(10)	GLB(5)	GLB(10)
Technology	IYW	0.95	0.89	0.82	0.86	1.00	13.04*	21.00*	2.16	11.41
		(-1.52)	(-1.41)	(-1.43)	(-0.73)	(0.64)	(0.02)	(0.02)	(0.83)	(0.33)
	XLK	0.90*	0.81*	0.74	0.77	1.00	34.57**	47.51**	4.44	12.26
		(-2.40)	(-2.05)	(-1.82)	(-1.06)	(0.68)	(0.00)	(0.00)	(0.49)	(0.27)
	VGT	0.95	0.88	0.82	0.86	1.00	12.93*	20.32*	1.90	11.39
		(-1.36)	(-1.42)	(-1.42)	(-0.71)	(0.73)	(0.02)	(0.03)	(0.86)	(0.33)
Telecommunication	IYZ	0.97	0.88	0.80	0.83	1.00	15.30**	29.81**	6.57	12.37
		(-0.81)	(-1.16)	(-1.23)	(-0.71)	(0.53)	(0.01)	(0.00)	(0.25)	(0.26)
	XTL	0.95	0.93	0.86	0.88	0.99	15.18**	24.28**	8.68	14.05
		(-0.95)	(-0.56)	(-0.74)	(-0.44)	(0.39)	(0.01)	(0.01)	(0.12)	(0.17)
	VOX	0.97	0.87	0.78	0.77	1.00	15.90**	24.90**	4.25	7.31
		(-0.72)	(-1.31)	(-1.38)	(-0.99)	(0.57)	(0.01)	(0.01)	(0.51)	(0.70)
Utility	IDU	0.91*	0.84	0.79	0.76	1.00	29.07**	53.91**	2.86	3.87
		(-2.02)	(-1.53)	(-1.26)	(-0.97)	(0.55)	(0.00)	(0.00)	(0.72)	(0.95)
	XLU	0.89*	0.82	0.75	0.72	1.00	38.62**	87.62**	2.13	4.44
		(-2.30)	(-1.70)	(-1.47)	(-1.15)	(0.45)	(0.00)	(0.00)	(0.83)	(0.93)
	VPU	0.90*	0.82	0.76	0.73	1.00	35.15**	65.09**	2.53	3.59
		(-2.13)	(-1.63)	(-1.38)	(-1.07)	(0.53)	(0.00)	(0.00)	(0.77)	(0.96)

Table 1.5: The test statistics of 9 ETFs which track major US equity sector indices for the period Jan 1, 2006 - Sep 30, 2013. The variance ratios $1 + M(q)$, $q = 2, 5, 10, 20$, are reported in VR labeled columns and in the main rows in line with the ETF tickers, whereas the corresponding z-scores are shown in parentheses immediately below each main row. The DF labeled column shows the first lag regression coefficients of log prices for the Dickey-Fuller test in the main rows with the corresponding p-values in parentheses immediately below the main rows. The LB labeled columns represent the Q-statistics for the Ljung-Box test on the innovations w_k in the main rows with p-values in parentheses immediately below each main row. Similarly, the GLB labeled columns show the Q-statistics for the Ljung-Box test on the standardized residual e_k from fitting returns to a GARCH(1,1) model.

have considerable impacts; the corresponding variance ratios are significant up to over 10 lags. Group III has few significant test statistics, supporting the random walk hypothesis. In particular, the material sector ETFs show none rejections, which exhibit a strong similarity with the commodity ETFs investigated in later sections. Group II ETFs have very low p-values for all Q-statistics in the LB columns; several show considerable rejections in the variance ratio tests such as the utility sector. The information contents of the ETFs in this group, however, are mostly removed after fitting to the GARCH model.

Many test statistics behaviors observed before remain valid. For instance, all variance ratios for sector ETFs are less than 1 across different lags except for one case in the consumer discretionary sector, where VCR has $M(2) + 1 = 1.01$. Variance ratios and their absolute z-scores typically decrease in q for $q \leq 10$. All Dickey-Fuller tests fail to reject the existence of unit roots in the log price dynamics.

Sector performances tend to be driven by their own characteristics and economic fundamentals. Conover, Jensen, Johnson and Mercer [Conover *et al.* (2008)] investigate the differences among sector performances by understanding how incorporating monetary policy signals can improve the sector rotation strategy. They divide monetary policies into two classes, expansive and restrictive, and categorize sectors by cyclical or defensive. Cyclical sectors tend to be more sensitive to market changes than defensive ones, exhibiting higher than average beta. For instance, stocks in the Consumer Discretionary sector are of this kind. They consist of businesses that sell nonessential goods and services like media products, automobiles, consumer apparels, etc., and are historically out-performers in early recovery phase of a business cycle. Defensive sectors are the opposite and remain stable even in recession. Consumer Staples sector is such an example. They consist of companies that sell necessities like food and drugs. As a result, the Consumer Discretionary section shows more trending behavior than the Consumer Staples, exhibiting relatively higher autocorrelation.

REIT ETFs stand out from our analysis by having the lowest levels of variance ratios with the strongest significance. This however is quite different from what has been observed before. Quite a few authors find that in periods prior 2006 REITs' returns are often characterized by positive autocorrelations and greater-than-one variance ratios ([Jirasakuldech and Knight (2005)], [Kuhle and Alvayay (2000)], [Wang *et al.* (1995)], [Han (1991)]). Some of them argue that this is because new information in REITs is not made available to all investors at the same time and even if it is, many REIT investors react rather slowly to the arrival of new information. Others show that when compared with the general stock market, REIT stocks tend to have a lower level of institutional investor participation and are followed by fewer security analysts. Furthermore, REITs that have a higher percentage of institutional investors or are followed by more security analysts tend to perform better than other REITs. Therefore, under-reaction has led to rejecting the random walk null hypothesis in those earlier time periods. Since 1992, the REIT market capitalization has gained an average annual growth of over 25%. More investors start to consider REITs as a solid income-generating investment. The introduction of the REIT ETFs not only reflects the REITs' popularity but also makes it easily accessible to an even wider pool of investors. The fact that the variance ratios has turned to levels near 0.8 for the period January 2006 - September 2013 may be associated with the situation where illiquidity no longer dominates the performance of REITs. In addition, since the housing bubble bursted during the Subprime Mortgage Financial Crisis, the real estate market has been considerably undervalued compared to its fundamentals and historical levels. This undervaluation together with REIT's growth is likely to explain the negative serial correlation in REIT ETFs and indices for the most recent decade.

For completeness of our investigation on REIT ETFs, we also perform variance ratio tests for the period December 1995 - December 2005. Due to REIT ETFs' lack of data points before 2006, we use the MSCI US REIT Index instead. We expect this to offer reasonably consistent results because the REIT ETF VNQ tracks this MSCI index and since the inception of VNQ, it has tracked the index rather closely. Table 1.6 summarizes the corresponding test statistics. All of the variance ratios reported appear greater than 1, statistically significant enough to reject the random walk hypothesis with over 99% confidence. In particular, REIT index returns show significant positive serial correlation at a level of 0.18 with over 99% confidence.

One additional interesting pattern we have observed from the three

q	2	5	10	20
Variance Ratio	1.18**	1.38**	1.36**	1.26
z-score	(5.31)	(5.28)	(3.43)	(1.78)

Table 1.6: The variance ratio statistics on the MSCI US REIT Index for the period January 1997 - December 2005.

tables is that the State Street SPDR sector ETFs [9] tend to have variance ratios of higher significance compared to the iShares and Vanguard ETFs. For instance, among the ETFs from the Energy and Technology sectors listed in Tables 1.3, 1.4 and 1.5, XLE and XLK are the only ETFs that have significant test statistics. We think that this is not surprising and can be justified by our results in the earlier subsection on size and style that large-cap and value stocks tend to behave less like random walks. Statistics on the compositions of the SPDR sector ETFs show that they mostly consist of large-cap companies, whereas iShares and Vanguard sector ETFs account for significant portions of small-cap and mid-cap assets. Take the year of 2013 as an example, XLK has almost 90% weight in large-cap and giant-cap stocks, but only 10.83% in mid-caps and 0.09% in small-caps. In contrast, IYW has 14.72% in mid-caps and 3.47% in small caps and VGT has 15.88% in mid-caps, 5.52% in small-caps and 2.09% in micro-caps.

1.3.1.4 *Effects of size and style on sector ETFs*

To understand how size and style affect sector performances more carefully, we design a style score and assign it to each of the 32 sector ETFs. Then, we plot the scores (x-axis) versus the market capitalizations (y-axis) for all these ETFs as a two-dimensional scattered chart in Figure 1.3. The score measures the style of an ETF and is derived from two fundamentals, dividend yield and book-to-price ratio. By definition, growth stocks reinvest most of their dividends and tend to be overvalued, therefore, we expect both statistics to be low for growth stocks and high for value stocks. We collect these ratios from Morningstar's website as of Oct 11, 2013 and report them in Table 1.7 together with the proposed style scores and market capitalizations in billion dollars.

[9]SPDR sector ETFs have tickers that start with the letter X, iShares ETFs start with the letter I and Vanguard ETFs start with the letter V.

ETF Tickers	BP	DIV	Score	Market Capitalization
IYC	0.29	1.34	1.81	34.02
XLY	0.27	1.43	1.36	40.20
VCR	0.29	1.36	2.11	23.19
IYK	0.28	2.48	18.37	38.74
XLP	0.29	2.84	24.27	65.77
VDC	0.29	2.59	20.29	53.83
IYE	0.56	2.08	34.07	62.50
XLE	0.56	2.04	33.72	58.86
VDE	0.57	2.06	34.78	57.02
IYF	0.83	2.13	56.04	35.19
XLF	0.90	1.88	58.16	56.03
VFH	0.83	2.32	59.45	26.80
IYH	0.35	1.79	13.05	47.66
XLV	0.35	1.92	15.11	59.62
VHT	0.34	1.66	10.61	38.48
IYJ	0.34	1.85	13.02	25.00
XLI	0.32	2.10	15.52	40.33
VIS	0.37	1.90	16.12	23.45
IYM	0.42	2.66	32.15	19.19
XLB	0.38	2.79	30.44	24.50
VAW	0.42	2.46	28.85	13.17
IYR	0.50	4.65	68.25	9.41
VNQ	0.45	3.63	48.76	8.26
IYW	0.36	1.75	13.23	80.00
XLK	0.34	2.26	19.86	100.49
VGT	0.35	1.39	7.20	55.74
IYZ	0.51	3.48	51.39	10.03
XTL	0.52	1.53	22.77	4.22
VOX	0.45	3.63	48.92	15.52
IDU	0.64	4.11	71.26	14.40
XLU	0.66	4.34	76.42	18.68
VPU	0.61	3.87	65.16	13.96

Table 1.7: Book-to-price ratios and dividend payout ratios for the 32 sector ETFs retrieved from the Morningstart.com as of Oct 11, 2013. The first column represents the tickers of the ETFs, the second column book-to-price ratios, the third column dividend yield in percentage, the fourth column the computed style scores and the last column reports the ETFs' market capitalizations in billion dollars.

For each statistic, we have 32 data points. We denote the dividend payout ratio for the ith ETF by DIV_i and book-to-price ratio BP_i, $i = 1, 2, ..., 32$. We normalize them by defining

$$\widehat{DIV}_i = \frac{DIV_i - \min_k DIV_k}{\max_k DIV_k - \min_k DIV_k} \times 100 \tag{1.32}$$

and

$$\widehat{BP}_i = \frac{BP_k - \min_k BP_k}{\max_k BP_k - \min_k BP_k} \times 100 \tag{1.33}$$

so that the normalized statistics both range from 0 to 100. Then, the score we assign to each ETF is defined as the average of these two normalized statistics, that is,

$$Score_i = \frac{\widehat{BP}_i + \widehat{DIV}_i}{2}, \tag{1.34}$$

which again ranges from 0 to 100. The closer the score is to 100, the ETF is more likely to be of the value type; whereas, the closer the score is to 0, the ETF is more likely to be of the growth type.

By comparing this figure with Tables 1.3, 1.4 and 1.5, we see that the sector ETFs which have lower variance statistics and higher significance tend to locate far away from the origin (the left bottom corner of the figure). For instance, the REIT ETFs IYR and VNQ are near the bottom right frontier of the figure and the financials ETFs are near the top right frontier. To make this relationship easier to visualize, we separate the 32 ETFs into three groups based on the maximum of the absolute value of their z-scores over lags $q = 2, 5, 10, 20$. The higher the score, the more negative the autocorrelation. Group 1 consists of ETFs whose maximal absolute z-scores across various q's is greater than 1.5 and labeled by circles filled circles. Group 2 consists of ETFs whose maximal absolute z-scores across various q's is between 1 and 1.5 and is labeled by squares. Group 3 consists of ETFs whose maximal absolute z-scores across various q's is less than 1 and is labeled by diamonds. The clustering of the markers of these groups conform with our earlier understanding in the section that size and style are important factors influencing the level of autocorrelation of ETFs returns, despite the idiosyncratic sector characteristics.

In addition, we perform regression analysis on the z-scores of the sector ETFs' variance ratios at lag 2 and 5 against their market capitalizations and style factors. This will validate our observations in Figure 1.3 statistically.

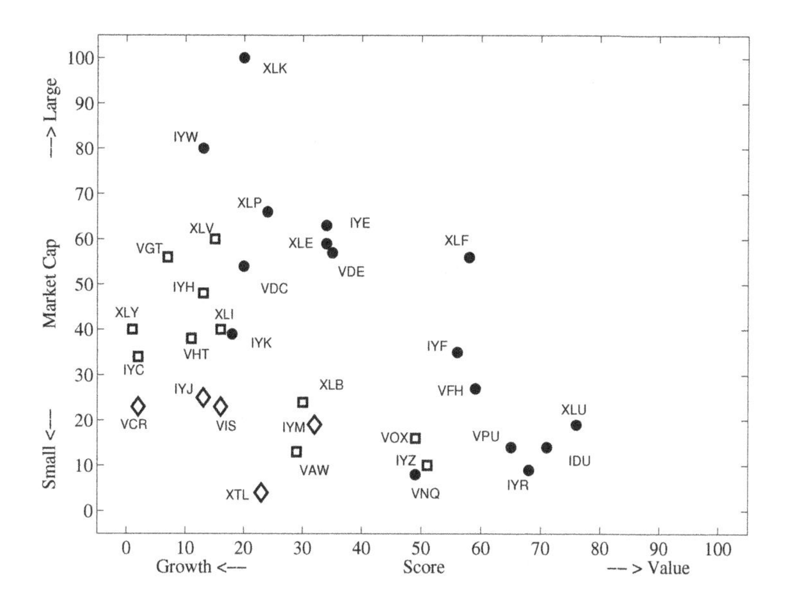

Fig. 1.3: The sector ETFs by its style score (x-axis) and the market capitalization (y-axis). The style of an ETF is characterized by the *Score* variable, whereas the size of an ETF is by its market capitalization in billion dollars. The closer the score is to 100, the style of the ETF is more likely to be value; the closer the score is to 0, the style of the ETF is more likely to be growth. The filled circles represent ETFs (Group 1) with low variance ratios and high significance. The squares represent ETFs (Group 2) with medium variance ratios and medium significance. The diamonds represent ETFs (Group 3) with high variance ratios and low significance.

Denote the z-score for the ith ETF at lag q by $Z_{i,q}$ and the corresponding market capitalization by $Size_i$. Then, the regressions we perform will be of the form

$$Z_{i,q} = \alpha + \beta \; Score_i + \gamma \; Size_i. \tag{1.35}$$

We also consider two more "nested" specifications where β and γ are set at 0 respectively, that is,

$$Z_{i,q} = \alpha + \beta \; Score_i \tag{1.36}$$

and

$$Z_{i,q} = \alpha + \gamma \; Size_i. \tag{1.37}$$

By comparing these three models, we will be able to understand the explanatory powers of Score and Size factors both individually and jointly.

One issue may occur in this line of investigation. In regression analysis, the test statistics often look better when more explanatory variables are added. For instance, the R^2 statistics have been shown to be a non-decreasing function of the number of independent variables in the corresponding regression equation. To keep consistency in comparing nested regression specifications we are considering, we perform additionally an F-test to check whether including one more variable in the regression will explain more variability in the dependent variable in a robust manner. Suppose we have two regression models, where model 1 is nested in model 2. Then, the test statistic will be of the form

$$F(p_2 - p_1, n - p_2) = \frac{(SSR_1 - SSR_2)/(p_2 - p_1)}{SSR_2/(n - p_2)}, \tag{1.38}$$

where SSR_i is the sum of squared residuals of model i, n is the total number of data points and p_i is the number of parameters in model i. It intuitively describes the additional variability explained in model 2 over model 1 as a "percentage" of the total variability explained by model 2. This statistic follows a F distribution with degrees of freedoms $(p_2 - p_1, n - p_2)$.

All test results for the regression analysis have been summarized in Table 1.8. It appears that the style score and market capitalization are statistically significant factors in explaining the variability of the variance ratios' z-scores for sector ETFs; their explaining power is the strongest when the two factors are used together. For instance, the coefficients β and γ have the most confidence in the specification Score and Size, where the adjusted R^2's are the largest among the three specifications, as well. All the F-tests give rather small p-values, suggesting that the third specification is superior. The regression coefficients, however, are negative in nature in all regression specifications considered. This says that the z-scores of the variance ratios decrease in Score and Size factors. Since all variance ratios are negative for sector ETFs, this means that the absolute magnitudes of their z-scores tend to be larger for sector ETFs with larger dividend, book-to-market ratios and market capitalization. Therefore, large-cap value (or liquid undervalued) ETFs tend to have smaller variance ratios statistics with stronger significance or their log-returns have stronger negative serial correlation.

Specification	α	β	γ	Adjusted R^2	SSR	F statistic
q=2						
Score	-0.961	-0.022		0.28	16.76	10.218
	(-4.13)	(-3.62)				(0.003)
Size	-1.423		-0.006	0.00	23.41	25.800
	(-4.82)		(-0.93)			(0.000)
Score and Size	-0.079	-0.029	-0.018	0.45	12.39	
	(-0.23)	(-5.07)	(-3.20)			
q=5						
Score	-1.143	-0.014		0.21	9.19	9.849
	(-6.63)	(-3.05)				(0.004)
Size	-1.374		-0.005	0.01	11.54	19.771
	(-6.63)		(-1.15)			(0.000)
Score and Size	-0.499	-0.019	-0.013	0.39	6.86	
	(-1.95)	(-4.44)	(-3.14)			

Table 1.8: The regression of the z-scores of variance ratios of sector ETFs on their style scores and market capitalization as in Eq. (1.35). The upper half refers to lag $q = 2$, whereas the lower refers to lag $q = 5$. The regression specification is in the first column, where it shows the names of the independent variables used to explain the dependent variable. In the columns labeled by α, β and γ, the values of the corresponding coefficients are displayed in the main rows in line with the regression specification; their t-statistics are displayed in parentheses immediately below the main rows. Adjusted R^2 and sum of squared regression residuals (SSR) are reported in the next two columns. The last column of this table shows the F-statistics for the nested regression tests in the main rows and their p-values in parentheses immediately below the main rows.

1.3.1.5 *Equity ETFs and their tracking indices*

So far in our analysis, we have taken advantage of the existing diversification of the equity ETF market, that many sub-categories of the broad market ETFs have already been created (such as sector ETFs). However, there remain some concerns. One of them is that many of the ETFs we considered have rather limited history and it is not easy to compare with previous research. Existing literature on the autocorrelation of the daily returns for broad market indices or equity portfolios can often be found in the 1990s and their reported historical levels of autocorrelation were strongly positive. For instance, Campbell, Grossman and Wang [Campbell *et al.* (1993)] documented that the autocorrelation of the daily returns for the value-weighted portfolio of all CRSP database stocks had an autocorrelation of 0.219 for the period 1962-1987. However, the autocorrelations we observed for the equity index ETFs in the period 2006-2013 are all negative (the lag-2 variance ratios are all less than unity). Therefore, it is necessary to validate if those ETFs' underlying indices had similar levels of autocorrelations for the same period. If so, then we may extend the history of equity ETFs by their underlying indices or portfolios, and thereafter study how their return autocorrelations evolve over time.

In Table 1.9, we report the sample 1-lag autocorrelations of daily returns of the equity ETFs we have already investigated for the period 2006-2013 and compare them with the autocorrelations for their tracking equity indices for the same period[10]. We also include the ETF's index tracking error in the last column of the table. The statistics in the table show that the indices' autocorrelations tend to be slightly lower than those of the ETFs by 0.01 on average. However, they all fall below zero. Some even reached the level of -0.2. This is strikingly different from the price return behaviors in 1980s and 1990s. For instance, in the period 1985-1995, the first lag daily autocorrelation of the S&P 500 Index is 0.03, the one of the Russell 1000 Index is 0.08 and the one of the Russell 2000 Index is 0.30. In addition, we also observe that the tracking errors computed using daily returns are rather small as well with their average being 30 basis point.

The fact that the serial correlations between those ETFs and their tracking indices do not significantly differ suggests we can understand factors which affect the levels of the autocorrelations in equity ETFs by researching the autocorrelations of equity portfolios. It also implies that we can easily get exposure to large portfolio of individual stocks by trading the corresponding index ETFs. This offers ETFs considerable advantage in comparison with the traditional index mutual funds, because ETFs not only offers intraday liquidity but also requires no minimum investments.

[10]ETFs whose underlying indices do not have sufficient data histories were omitted in Table 1.9.

Ticker	Index	ETF	Index-ETF	Tracking Error	Ticker	Index	ETF	Index-ETF	Tracking Error
IYR	-0.22	-0.19	-0.035	0.0042	IWS	-0.08	-0.06	-0.018	0.0018
VNQ	-0.22	-0.20	-0.014	0.0033	IYW	-0.08	-0.05	-0.023	0.0017
IYF	-0.15	-0.11	-0.038	0.0030	VGT	-0.08	-0.05	-0.028	0.0017
VFH	-0.14	-0.12	-0.028	0.0038	IJJ	-0.08	-0.07	-0.004	0.0016
XLF	-0.14	-0.12	-0.015	0.0040	IYK	-0.08	-0.08	-0.001	0.0019
IWN	-0.13	-0.11	-0.016	0.0026	VHT	-0.07	-0.06	-0.010	0.0014
XLU	-0.12	-0.11	-0.015	0.0027	IYH	-0.07	-0.06	-0.010	0.0014
IVW	-0.12	-0.10	-0.017	0.0018	IJT	-0.07	-0.06	-0.009	0.0017
SPY	-0.12	-0.09	-0.024	0.0021	IWO	-0.07	-0.06	-0.003	0.0026
VPU	-0.11	-0.10	-0.013	0.0013	IWR	-0.06	-0.04	-0.021	0.0017
IWD	-0.11	-0.10	-0.010	0.0021	MDY	-0.06	-0.06	0.006	0.0027
IDU	-0.11	-0.09	-0.021	0.0017	IYC	-0.05	-0.04	-0.017	0.0014
IJS	-0.11	-0.10	-0.010	0.0018	XLI	-0.05	-0.05	-0.006	0.0033
IYE	-0.11	-0.13	0.026	0.0037	XLB	-0.05	-0.04	-0.016	0.0036
IVE	-0.11	-0.09	-0.018	0.0016	VOX	-0.05	-0.03	-0.019	0.0050
IWB	-0.10	-0.09	-0.019	0.0016	VIS	-0.04	-0.03	-0.013	0.0014
IWM	-0.10	-0.09	-0.013	0.0024	IYJ	-0.04	-0.03	-0.015	0.0012
XLE	-0.10	-0.09	-0.008	0.0029	VAW	-0.04	-0.03	-0.015	0.0015
VDE	-0.10	-0.08	-0.012	0.0018	IYM	-0.04	-0.02	-0.019	0.0017
IWF	-0.09	-0.07	-0.021	0.0020	IWP	-0.04	-0.04	-0.004	0.0023
SLY	-0.09	-0.09	0.001	0.0125	IYZ	-0.04	-0.04	-0.001	0.0026
XLK	-0.09	-0.10	0.008	0.0029	IJK	-0.03	-0.02	-0.009	0.0015
XLV	-0.09	-0.08	-0.011	0.0033	VCR	-0.02	0.01	-0.024	0.0014

Table 1.9: The comparison of the autocorrelations of equity ETFs with those of the ETFs' tracking indices for the period 2006-2013. The "Index" column contains the first lag autocorrelation of ETF's tracking index daily return. The "ETF" column reports the corresponding ETF's autocorrelation using its daily returns. The "Index-ETF" column shows the difference between the index autocorrelation and the ETF autocorrelation. The last column shows the ETF's tracking error.

1.3.1.6 *Equity index autocorrelations and interest rates*

To see how levels of autocorrelation evolve over time, we add the 3 equally-weighted Fama-French size portfolios to our research. Compared to ETFs, it offers a much longer data history ranging from 1926 to 2013. We compute the first lag autocorrelations of daily returns rolling over the entire history with a window of 5 years for the portfolio of small-cap stocks, mid-cap stocks and large-cap stocks respectively. The upper panel of Figure 1.4 plots the corresponding rolling serial correlations.

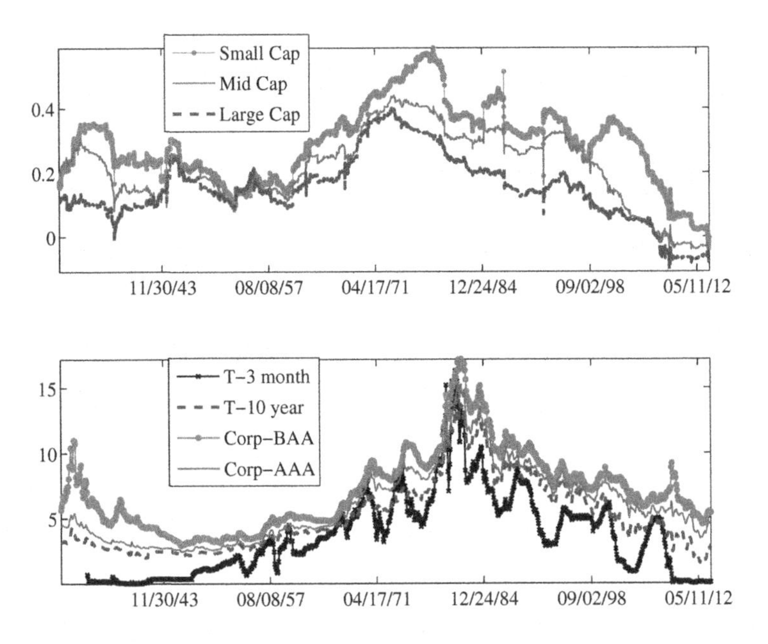

Fig. 1.4: The upper panel plots the first lag autocorrelations of daily returns rolling over the entire history with a window of 5 years for the portfolio of small-cap stocks, mid-cap stocks and large-cap stocks respectively for the period 1926-2013. The size portfolios are equally weighted. The lower panel plots the interest rates level of Treasuries over 10 year maturities, Treasuries of 3-month maturities, corporate bonds with the AAA rating and corporate bonds with the BAA rating for the same period.

We see that the levels of autocorrelation have experienced tremendous ups and downs within the sample period. It appears that they are related to the economic cycles, for which we have also charted the interest rate levels of Treasuries over 10 year maturities, Treasuries of 3 months, corporate bonds with AAA rating and corporate bonds with BAA rating in the same period for comparison in the lower panel of Figure 1.4. From this figure, we see a positive relationship between the levels of autocorrelation of broad market indices and those of interest rates, that is, strongly positive autocorrelations of equity indices tend to occur in environments with high interest rates, whereas significant levels of negative autocorrelations of equity indices tend to take place when interest rates are close to zero. From an economics perspective, a low interest rate policy leads to high prices of bonds and encourages investments in the stock market. It is a signal which indicates that the equity markets are undervalued. For instance, suppose we adopt the discounted cash flow or the dividend model to evaluate the intrinsic value of a stock. When the Fed lowers the rates, it practically implies that the present value of all the future cash flows of this company should be higher. As a result, the current market prices of the company are relatively low to its fundamentals. Hence, the stock is undervalued.

To justify this observation in a statistical manner, we compute the autocorrelation (denoted by ρ) of daily returns of the Fama-French portfolios based on sizes for each two consecutive years and regress them on the average levels of the rates (denoted by r) for the same period in the following form

$$\rho = c_1 + c_2 \log(r). \tag{1.39}$$

We illustrate the results via the regression statistics of the small cap portfolio versus AAA rated corporate bond rates in Table 1.10. The other pairs of portfolios and rates offer similar relationship, whose regression statistics are therefore omitted for brevity.

The table summarizes the regression statistics of the autocorrelation of the returns for small cap Fama-French portfolio on AAA rated corporate bond rates for every two consecutive years of three periods: 1931-1980, 1981-2013 and 1931-2013. In the table, each period corresponds two rows of data. The first row shows the estimated coefficients with their t-stats immediately below in parentheses. We see that the estimated coefficients for c_2 are statistically significant for all three periods. The positive sign of the coefficient also confirms with the empirically observed positive relationship between portfolio autocorrelations and interest rates.

Period	c_1	c_2	Adjusted R^2
1931-2013	-0.0077	0.1704**	0.31
	(-0.16)	(6.15)	
1931-1980	-0.0879**	0.2652**	0.67
	(-2.14)	(9.80)	
1981-2013	-0.5691**	0.4104**	0.77
	(-7.04)	(10.24)	

Table 1.10: The regression statistics of the autocorrelation of the returns for small cap Fama-French portfolio on AAA rated corporate bond rates for every two consecutive years of three periods: 1931-1980, 1981-2013 and 1931-2013. For each period, the first row shows the estimated coefficients with their t-stats immediately below in parentheses. The regression is of the form $\rho = c_1 + c_2 \log(r)$, where ρ represents the first lag autocorrelation and r represents the interest rates.

From Figure 1.4, we also observe the robust size effects on the first lag return autocorrelation. Portfolios of small caps tend to show higher autocorrelation than that of portfolios of large caps. The size of a stock is a natural measurement of liquidity for the following reasons: (1) Large caps have relatively more shares outstanding; they have significantly more trading volumes and often have smaller spreads than small caps. Small caps tend to show stale prices more frequently. (2) Information on small-cap stocks is less transparent and less available to the majority of investors compared with large caps. Typically small-cap companies have less analyst coverage. From our results above, we may conclude that portfolios based on less liquid stocks tend to have higher serial correlation[11].

With the observation of the effect of interest rates (or financing costs) in Figure 1.4, we may extend Lo and MacKinlay's non-synchronous trading model to better explain the portfolio index autocorrelation magnitudes and its dynamics. Lo and MacKinlay proposed a simple model and attempted to justify the strong positive autocorrelation observed in equity portfolios, especially for small-cap portfolios, for the period 1962-1994. However, they

[11] It is worth noting that there has been research which shows that the effects of liquidity measured by bid-ask spread and non-synchronous trading are the opposite between individual stocks and portfolio of stocks [Lo and MacKinlay (1988)] [Lo and MacKinlay (1990)]. In here, we are interested in portfolios of assets or portfolio-based indices and ETFs.

found non-synchronous trading alone was not able to explain entirely the portfolio return autocorrelation. For instance, between 1962 and 1994, the small-cap portfolio showed an empirical non-trading probability of 0.225, yet their model implied a non-trading probability of 0.394. For the large-cap portfolio, the empirical non-trading probability was 0.002, but the model suggested 0.188.

The basic assumption of their model is that the return of an individual security i, r_{it}, at time t can be represented by a one-factor model

$$r_{it} = \mu_i + \beta_i f_t + \epsilon_{it}, \tag{1.40}$$

where f_t and ϵ_{it} are assumed to be uncorrelated both temporally and cross-sectionally. If financing costs (interest rates) are indeed a factor of equity index autocorrelation, then we may incorporate this into the one-factor model by assuming that the factor f_t is autocorrelated as in

$$f_t = \alpha f_{t-1} + w_t, \tag{1.41}$$

where w_t is iid noise and is also independent from the idiosyncratic innovation ϵ_{it}'s. The sign and magnitude of α will vary according to interest rates. Then, in Chapter 2, we show that a well-diversified portfolio in turn will carry a portion of α in its autocorrelation, that is, the autocorrelation of the index will be the sum of the pure non-synchronous trading effect and a multiple of α. The non-synchronous effect will help justify the autocorrelation gap between small-cap index and large-cap index, whereas the multipler of α will explain the variation of magnitudes of index autocorrelation over time as we have seen in Figure 1.4.

1.3.1.7 *Effects of volatility on index autocorrelation*

LeBaron (1992) found that serial correlations of equity index returns are inversely related to their variances. Therefore, we added this factor into the regression model so that it becomes

$$\rho = c_1 + c_2 \log(r) + c_3 \nu, \tag{1.42}$$

where ν is the variance of daily returns of the small-cap Fama-French portfolio computed using the same samples for ρ. The statistics are summarized in Table 1.11. It shows that the coefficient of the variance term is statistically significant and negative, indicating that there is an inverse relationship

Period	c_1	c_2	c_3	Adjusted R^2
1981-2013	-0.5691**	0.4104**	-	0.77
	(-7.04)	(10.24)	-	
1981-2013	-0.4241**	0.3552**	-11.9019**	0.81
	(-4.55)	(8.34)	(-2.58)	

Table 1.11: The regression statistics of the autocorrelation of the returns for small cap Fama-French portfolio on AAA rated corporate bond rates and annualized variances for every two consecutive years of the period 1981-2013. For each period, the first row shows the estimated coefficients with their t-stats immediately below in parentheses. The regression is of the form $\rho = c_1 + c_2 \log(r) + c_3\nu$, where ρ denotes autocorrelation, r denotes interest rates and ν denotes the annualized variances.

between variance and autocorrelation for the small cap portfolio. The adjusted R^2 increased from 0.77 to 0.81, adding to the robustness of the new control variable.

1.3.1.8 *100 Fama-French portfolios*

In this subsection, instead of using the well-defined sector ETFs to reveal the effects of size and style, we create our own sub-portfolios of the 100 equally weighted Fama-French portfolio and perform regression analysis on them. The data are sourced from the US research return data from the data library on the website[12] of Kenneth French which contains the daily returns, annual book-to-market ratios and market capitalizations of the 100 portfolios formed on size and book-to-market ratio based on all CRSP firms incorporated in the US and listed on the NYSE, AMEX or NASDAQ[13]. We will consider the sample period 1981-2013 and perform two experiments.

[12] *http://mba.tuck.dartmouth.edu/pages/faculty/ken.french/data$_l$ibrary.html*

[13] The total range of dates for the available daily return data is July 1, 1926 - January 30, 2014. The portfolios, which are constructed at the end of each June, are the intersections of 10 portfolios formed on size (market equity, ME) and 10 portfolios formed on the ratio of book equity to market equity (BE/ME). The size breakpoints for year t are the NYSE market equity deciles at the end of June of t. BE/ME for June of year t is the book equity for the last fiscal year end in $t-1$ divided by ME for December of $t-1$. The BE/ME breakpoints are NYSE deciles. Note β = ME/BE.

Experiment 1: group by BE/ME

We first regroup the 100 portfolios by BE/ME (book-to-market ratio) by adding the 10 portfolios with similar BE/ME but different ME (market capitalization). There are dates when several of the 100 portfolios do not have available data, we assume those return entries to be zero. The resultant 10 portfolios will be roughly of the same ME but different BE/ME at each point in time within the period of consideration. The reason we perform this modification is that (1) we would like to see more carefully how the book-to-market ratios affect portfolio daily autocorrelations when the average size of the portfolio is controlled, (2) some of the 100 Fama-French portfolios either lack of data or have too few number of consituents. Since the regrouping of portfolio returns is done daily, as we consider different periods, the corresponding controlled market capitalization of the 10 portfolios will change. So market capitalization may still show up as a valid factor.

Once these 10 portfolios are obtained, we compute the autocorrelation of their daily returns for every non-overlapping two years. At the same time, we also calculate the realized volatility and average interest rates[14] for the same period. We then perform regression of the autocorrelations of all 10 portfolios over all 2-year sub-periods on their average book-to-market ratios (β), the log of their average market capitalization, realized volatility and the log of the corresponding average interest rates. That is,

$$\rho = c_1 + c_2 \frac{1}{\beta} + c_3 \log(size) + c_4 vol + c_5 \log(r). \tag{1.43}$$

The statistics are summarized in Table 1.12 for five regression specifications. The first specification contains all four explanatory variables. The second specification removes the size factor from Specification 1. The next three specifications contain only two out of the three factors in Specification 2.

In Specification 1, we find all four factors to be statistically significant with an adjusted R^2 of 85.80%. The coefficients for book-to-price ratio, size and volatility are all negative, whereas the coefficient of interest rates is positive. This suggests that the daily autocorrelation of equity portfolios is larger when interest rates and the book-to-price ratios are larger

[14]AAA rated corporate bond rates are used for consistency with earlier analysis in this chapter

Specification	c_1	c_2	c_3	c_4	c_5	adjusted R^2
1	-0.0691	-0.0327	-0.0326	-0.5545	0.2750	0.8589
	(-0.38)	(-3.81)	(-2.40)	(-6.10)	(7.50)	
2	-0.4957	-0.0252		-0.5628	0.3549	0.8549
	(-13.64)	(-3.11)		(-6.11)	(23.08)	
3	-0.491			-0.5754	0.3436	0.8473
	(-13.19			(-6.10)	(22.42)	
4	-0.6488	-0.0274			0.4036	0.8232
	(-22.32)	(-3.07)			(27.80)	
5	0.3032	0.0192		-1.6629		0.3927
	(13.38)	(1.19)		(-10.32)		

Table 1.12: The regression statistics of the autocorrelation of the returns of 10 Fama-French portfolios based on BE/ME controlled by ME on BE/ME, ME, volatility and interest rates for the period 1981-2013, i.e. $\rho = c_1 + c_2\frac{1}{\beta} + c_3\log(size) + c_4 vol + c_5\log(r)$, where β denotes the book-to-price ratio. We show five nested specifications. For each specification, the first row shows the estimated coefficients with their t-stats immediately below in parentheses.

but when volatility and market capitalization of the portfolio are smaller. Therefore, portfolios of value stocks and large caps tend to have lower levels of autocorrelations.

Since these portfolios are constructed in control of the size factor, we do not expect the factor explains much of the variances in the dependent variable. Comparing Specifications 1 and 2, we find that the removal of the size factor does not reduce much of the adjusted R^2.

By comparing the Specifications 2-5, we find book-to-price ratio and volatility are somewhat correlated with one another. When the book-to-price ratio factor is removed from Specification 2, we find the coefficient of volatility becomes more negative; similarly, when the volatility term is removed from Specification 2, we find the book-to-price ratio coefficient becomes more negative. The interest rate term is shown to explain more of the temporal autocorrelation variation. When it is removed from Specification 2, the sign of the book-to-price ratio coefficient changes from negative to positive, trying to make up the explanation power of the interest rate term. To accommodate this alternation, the coefficient of volatility becomes significantly more negative. Nevertheless, the adjusted R^2 level was not recovered, realizing a drop from 82.32% to 39.27%.

Experiment 2: group by ME

In the second experiment, we perform essentially the same procedures except that we regroup the 100 portfolios using the ME (market capitalization) criterion. The resultant 10 portfolios will have similar levels of book-to-market ratios at each point of time within the sample period but different levels of market capitalization. Once these 10 portfolios are obtained, we perform the same regression. The corresponding statistics are summarized in Table 1.13 again with five specifications.

Specification	c_1	c_2	c_3	c_4	c_5	adjusted R^2
1	-0.0327	-0.0992	-0.0333	-0.7147	0.2731	0.8189
	(-0.60)	(-2.12)	(-11.15)	(-5.34)	(9.09)	
2	-0.0073		-0.0314	-0.8660	0.2240	0.8151
	(-0.14)		(-10.90)	(-7.58)	(11.62)	
3	-0.2136		-0.0342		0.2914	0.7526
	(-4.02)		(-10.35)		(14.73)	
4	-0.3190			-1.0249	0.2825	0.6847
	(-5.41)			(-6.93)	(11.68)	
5	0.5738		-0.0407	-1.4779		0.6668
	(22.78)		(-10.96)	(-10.87)		

Table 1.13: The regression statistics of the autocorrelation of the returns of 10 Fama-French portfolios based on ME controlled by BE/ME on BE/ME, ME, volatility and interest rates for the period 1981-2013, i.e. $\rho = c_1 + c_2\frac{1}{\beta} + c_3\log(size) + c_4 vol + c_5\log(r)$, where β denotes the book-to-price ratio. We show four nested specifications. For each specification, the first row shows the estimated coefficients with their t-stats immediately below in parentheses.

The first specification again contains all four explanatory variables. Specification 2 has the book-to-price ratio removed. The next three specifications have only two out of the three factors in Specification 2.

In Specification 1, all four factors are statistically significant with confidence over 99%. The adjusted R^2 is about 82%. Since the portfolios are constructed with the book-to-market ratio controlled so that when the corresponding factor is removed from Specification 1, the adjusted R^2 does not alter much.

From Specifications 2-5, we observe that all the three factors are robust, their signs and magnitudes do not change much when one is removed out of the three. In particular, the adjusted R^2 remain over 66% even when only two of them are present in the model, indicating the importance of these three factors. The qualitative results in this experiment are more or less consistent with the previous one. The first lag return autocorrelations of equity portfolios tend to be inversely associated with portfolio volatilities and positively with interest rates. Large autocorrelations often occur during low volatility and high interest rate environments. In addition, portfolios which are made up of large caps and value stocks tend to have smaller return autocorrelations.

1.3.2 *Bond ETFs*

Stocks are not the only instruments to which ETFs offer exposure. In July 2002, iShares made the first bond funds available, for instance. Among those ETFs, LQD[15] offers exposure to the liquid corporate bond market and SHY[16] offers exposure to the short-duration Treasury bonds with maturities ranging 1-3 years. By the end of 2006, there were about 6 fixed income ETFs available with approximately $20 billion in assets. By the end of 2008, the AUM of the market had almost tripled to over $56 billion.

Bond ETFs can be considered a rather impressive innovation, which attempts to put the OTC market on the exchange. Compared to stocks, the bond market is not as liquid or transparent. Bonds typically trade over the counter rather than on the exchanges. As a result, much of the bond trading is not done as liquidly as stocks. This feature of the bond market is inevitably reflected in the construction of bond ETFs. Unlike a typical index ETF, which generally consists of all constituent stocks. A bond fund often holds only a fraction of the bonds that make up the underlying index. Given the knowledge of bond prices and their relations with interest-rates,

[15]the iBoxx Investment Grade Corporate Bond ETF
[16]iShares Barclays Capital US 1-3 Year Treasury Bond Index Fund

coupons and maturity, bond fund managers may use a certain amount of technical sampling techniques to achieve a performance which is rather close to their tracking bond indices. For instance, the prospectus of the fund LQD states that it adopts a representative sampling indexing strategy, which involves investing in a representative sample of securities which collectively has an investment profile similar to that of the underlying index. The securities selected are expected to have investment characteristics (such as market capitalization and industry weightings), fundamental characteristics (such as return variation, duration, maturity or credit ratings) and liquidity measures similar to those of the underlying index. Generally, the fund invests over 90% of its assets in securities of the underlying index and at least 95% in investment-grade corporate bonds. However, the fund has the flexibility of investing in bonds not included in the underlying index but which its manager believes will help LQD track better the underlying index, such as repurchase agreements collateralized by US government obligations and in cash and cash equivalents, including shares of certain money market funds.

With the addition of bond ETFs as a choice on the exchange, investors become more capable of realizing their investment goals. Traditionally, investors had few options in trading fixed income securities. They were not liquid enough to participate in faster strategies, especially those with multi-asset exposures. Since not only the risk models need adjust to different trading environments, but the transaction costs require tailored calibrations as well. Target portfolios were generally harder to realize if investors want to trade quickly. The fact that bond ETFs are traded like stocks makes them much more flexible in participating strategies with various horizons, realizing investors' philosophies in an agile and effective way. Some market participants believe that compared to the mutual funds which offer similar exposure to the bond market, ETFs in this category are more efficient in adjusting their prices to bond-related market information.

The bond ETF market has achieved tremendous success since its inception, as well, especially during the 2007-2008 financial crisis. At the time, the trading volume in markets like corporate bonds declined drastically, primarily due to the fact that many of the firms were fighting to stay afloat. The bond ETFs however offered alternative exposure to the bond market with easy access, transparency and liquidity as simple as trading stocks on the exchange. Their AUM grew above \$50 billions in 2008 and reached almost \$150 billions by the end of 2010. In the second quarter of 2012, the global fixed income ETFs have sustained over \$250 billion of assets.

Figure 1.5 summarizes the total AUMs of all US fixed income ETFs since 2005. We see that in the year of 2013, the total assets managed by the US fixed income ETFs declined slightly. This reflects the fact that the US economy has gained strength. Comparing the observation here with Figure 1.1, we observe that money has overall flown from investments in bonds to stocks, as the assets managed by the equity ETFs in the US market hiked in the same year from just below \$1 trillion to almost \$1.4 trillion.

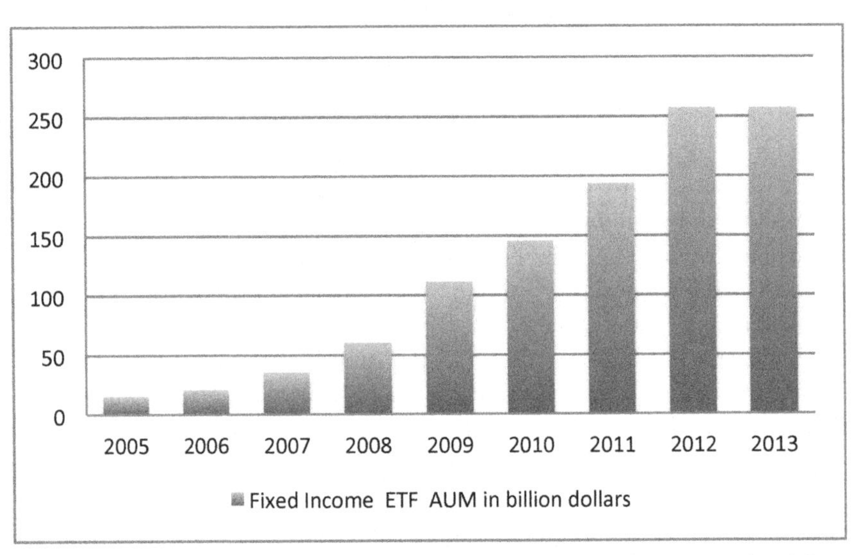

Fig. 1.5: The total AUMs of the fixed income ETFs in the US by year retrieved from Bloomberg.

In this section, we consider 8 fixed income ETFs, two for corporate bonds, five for treasury debts and 1 for mortgage. We find that during the economic environment with quantitative easing, the short-term Treasury debts based ETFs showed statistically significant mean-reversion behaviors and the corporate bond indices showed rather positive return autocorrelations compared to Treasuries based indices. Furthermore, the liquidity of the assets making up the bond indices and ETFs seems to have played a very important role for their return autocorrelations.

We apply the same methodologies as in earlier sections. The test statistics are summarized in Table 1.14 for the period January 2008 - September 2013. We see strong rejections of the random walk hypothesis for the short-term Treasury debt based ETFs, BIL and SHY, including the Dickey-Fuller tests. Some evidence from the Ljung-Box test indicates non-random-walk behaviors in LQD, HYG, IEI and MBB, but they do not have statistically significant variance ratios. The long-term Treasury bond based ETFs, IEF and TLT, passed all the tests we performed without rejection. All variance ratios are less than 1, except for HYG at lag 20. Their monotonicity in q varies among the funds considered. The GARCH models for the conditional variance of log-returns do not explain all information in the time series for HYG, BIL and SHY.

We next consider Treasury debt ETFs and non-Treasury debt ETFs separately.

1.3.2.1 *Treasury debt ETFs*

In Table 1.14, BIL, SHY, IEI, IEF and TLT represent the Treasury bills / bonds performances with various maturities covering 1-3 months, 1-3 years, 3-7 years, 7-10 years and 20+ years respectively. We observe that for the same level of q, the variance ratios of these ETFs (except TLT) increase as the maturity increases. For the same ETF, as q increases, the corresponding variance ratio tends to decrease, however, we do observe at the levels between 10 and 20 days, this relationship reverse for bonds with medium and long maturities, that is, monthly variances (annualized) for these bonds are larger than their bi-weekly variances (annualized).

Since $M(2)$ also represents the sample autocorrelation, we observe that the short-term Treasury bill / bond returns are negatively autocorrelated with strong confidence. For instance, the first lag return autocorrelation for BIL is -0.35 and that of SHY is -0.12 with t-stats -3.06 and -2.31 respectively.

Category	Ticker	VR(2)	VR(5)	VR(10)	VR(20)	DF	LB(5)	LB(10)	GLB(5)	GLB(10)
Bond	LQD	0.91	0.75	0.73	0.83	1.00	31.61**	184.99**	7.76	13.21
		(-0.54)	(-0.86)	(-0.76)	(-0.34)	(0.74)	(0.00)	(0.00)	(0.17)	(0.21)
	HYG	0.99	0.92	0.98	1.04	1.00	18.14**	82.74**	13.94*	16.25
		(-0.08)	(-0.45)	(-0.08)	(0.12)	(0.85)	(0.00)	(0.00)	(0.02)	(0.09)
	BIL	0.65**	0.33**	0.21*	0.17	0.97**	208.11**	244.96**	115.28**	128.80**
		(-3.06)	(-3.00)	(-2.41)	(-1.85)	(0.00)	(0.00)	(0.00)	(0.00)	(0.00)
	SHY	0.88*	0.73*	0.61*	0.60	1.00	30.39**	43.57**	17.26**	33.54**
		(-2.31)	(-2.38)	(-2.25)	(-1.66)	(0.06)	(0.00)	(0.00)	(0.00)	(0.00)
	IEI	0.95	0.87	0.78	0.84	1.00	15.25**	21.97*	8.05	13.08
		(-1.48)	(-1.60)	(-1.75)	(-0.91)	(0.33)	(0.01)	(0.02)	(0.15)	(0.22)
	IEF	0.97	0.94	0.86	0.94	1.00	8.70	14.00	6.63	12.07
		(-0.94)	(-0.88)	(-1.33)	(-0.40)	(0.50)	(0.12)	(0.17)	(0.25)	(0.28)
	TLT	0.96	0.91	0.83	0.90	1.00	8.54	13.22	6.62	13.19
		(-0.99)	(-1.19)	(-1.53)	(-0.64)	(0.45)	(0.13)	(0.21)	(0.25)	(0.21)
	MBB	0.93	0.96	0.71	0.75	1.00	32.78**	60.97**	5.82	9.25
		(-1.37)	(-0.37)	(-1.60)	(-0.98)	(0.29)	(0.00)	(0.00)	(0.32)	(0.51)

Table 1.14: The test statistics of 8 ETFs in the fixed income markets for the period Jan 1, 2008 - Sep 30, 2013. The variance ratios $1 + M(q)$, $q = 2, 5, 10, 20$, are reported in VR labeled columns and in the main rows in line with the ETF tickers, whereas the corresponding z-scores are shown in parentheses immediately below each main row. The DF labeled column shows the first lag regression coefficients of log prices for the Dickey-Fuller test in the main rows with the corresponding p-values in parentheses immediately below the main rows. The LB labeled columns represent the Q-statistics for the Ljung-Box test on the innovations w_k in the main rows with p-values in parentheses immediately below each main row. Similarly, the GLB labeled columns show the Q-statistics for the Ljung-Box test on the standardized residual e_k from fitting returns to a GARCH(1,1) model. Significance levels at 99% and 95% are signaled by two and one stars next to the main row statistics respectively. Under the random walk null hypothesis, the value of the variance ratio is 1, the Dickey-Fuller regression first lag coefficient is 1 and w_t's are independent. Under the random walk with GARCH residuals price hypothesis, the Dickey-Fuller regression first lag coefficient is 1, LB Q-statistics are significant and GLB Q-statistics are insignificant.

Bond prices are intimately related to the corresponding interest rate levels. We would like to see if interest rates of the corresponding maturities have similar behaviors with these Treasury debt based ETFs. We collect the interest rates historical data from the Federal Reserve's website (H. 15 releases[17]) and compute the various test statistics of these rates for the same sample period. The results are summarized in Table 1.15.

It appears that the essential behaviors of the variance ratios remain the same. Short-term interest rates returns are negatively serial correlated with strong confidence across all q levels considered. The variance ratios statistics are all significant for Treasuries with less than 2-year maturities. The Dickey-Fuller test reject unit-root hypothesis for 1-month and 3-month Treasury bills, offering strong evidence of stationarity and mean reversion. All Ljung-Box statistics have strong confidence for Treasuries with less than 2-year maturities. In addition, there is a clear transition of the efficiency in the Treasury securities at the maturity level of 3-years. All rates corresponding to Treasuries with longer maturities have strong evidence for the random walk like dynamics. It appears that for the same q level, the variance ratios for Treasury rates more or less increase with maturity reaching levels equal or slightly above 1 at the 30-year level, consistent with what we have seen from the corresponding ETFs. Similarly, the p-values for the Dickey-Fuller statistics increase with maturity. Therefore, although these Treasury ETFs trade like stocks, their price dynamics are considerably influenced by the underlying rates.

Factors affecting short-term and long term interest rates are usually different. Long-term rates are much more market driven, relying on the demand and supply for credit, inflation, etc., whereas short-term rates are largely monitored by the Federal Reserve. Roley (1982) shows that weekly money supply announcements have significant impacts on short-term interest rates' movements. Fleming and Remolona (1997) report the existence of striking intraday adjustment patterns for price volatility, trading volume, and bid-ask spreads in the US Treasuries market around the time of macroeconomic announcements. For the years 2008-2013, the Federal Reserve's main efforts and policies are to maintain a low interest rate environment, facilitating the economic recovery from the subprime mortgage financial crisis. It is likely that the unprecedented quantitative easing mea-

[17] http://research.stlouisfed.org/fred2/.

Category	Maturity	VR(2)	VR(5)	VR(10)	VR(20)	DF	LB(5)	LB(10)	GLB(5)	GLB(10)
Treasury	1-month	0.91*	0.58**	0.43**	0.30**	0.95**	76.55**	77.76**	37.68**	41.71**
Rate		(-2.23)	(-4.64)	(-4.18)	(-3.55)	(0.00)	(0.00)	(0.00)	(0.00)	(0.00)
	3-month	0.79*	0.43**	0.34**	0.28*	0.97**	107.85**	115.88**	18.99**	37.70**
		(-2.22)	(-3.13)	(-2.77)	(-2.44)	(0.00)	(0.00)	(0.00)	(0.00)	(0.00)
	6-month	0.86**	0.62**	0.50**	0.44*	0.99	49.95**	87.50**	26.61**	41.60**
		(-2.67)	(-3.10)	(-2.73)	(-2.23)	(0.07)	(0.00)	(0.00)	(0.00)	(0.00)
	1-year	0.86**	0.74**	0.67**	0.60*	1.00	29.51**	38.30**	23.09**	25.97**
		(-3.84)	(-3.32)	(-2.81)	(-2.34)	(0.17)	(0.00)	(0.00)	(0.00)	(0.00)
	2-year	0.88**	0.75**	0.65**	0.60**	0.99	31.88**	38.44**	27.16**	36.22**
		(-3.66)	(-3.52)	(-3.28)	(-2.61)	(0.18)	(0.00)	(0.00)	(0.00)	(0.00)
	3-year	0.97	0.91	0.79	0.79	1.00	8.58	19.21*	6.38	15.19
		(-0.85)	(-1.31)	(-1.95)	(-1.34)	(0.27)	(0.13)	(0.04)	(0.27)	(0.13)
	5-year	0.96	0.91	0.82	0.86	1.00	6.43	10.15	3.83	11.43
		(-1.22)	(-1.23)	(-1.67)	(-0.92)	(0.33)	(0.27)	(0.43)	(0.57)	(0.33)
	7-year	1.00	0.97	0.86	0.89	1.00	7.90	10.46	4.20	10.73
		(-0.14)	(-0.48)	(-1.28)	(-0.69)	(0.36)	(0.16)	(0.40)	(0.52)	(0.38)
	10-year	1.01	0.99	0.88	0.95	1.00	8.80	16.25	3.58	11.99
		(0.41)	(-0.18)	(-1.10)	(-0.29)	(0.35)	(0.12)	(0.09)	(0.61)	(0.29)
	20-year	1.01	0.97	0.88	0.95	1.00	6.59	13.48	2.21	10.18
		(0.41)	(-0.34)	(-1.12)	(-0.30)	(0.37)	(0.25)	(0.20)	(0.82)	(0.43)
	30-year	1.03	1.00	0.91	1.00	0.99	9.15	17.64	2.98	12.70
		(0.83)	(-0.05)	(-0.79)	(0.02)	(0.28)	(0.10)	(0.06)	(0.70)	(0.24)

Table 1.15: The test statistics of Treasury debt interest rates for the period Jan 1, 2008 - Sep 30, 2013 for various maturities. The variance ratios $1 + M(q)$, $q = 2, 5, 10, 20$, are reported in VR labeled columns and in the main rows in line with the ETF tickers, whereas the corresponding z-scores are shown in parentheses immediately below each main row. The DF labeled column shows the first lag regression coefficients of log prices for the Dickey-Fuller test in the main rows with the corresponding p-values in parentheses immediately below the main rows. The LB labeled columns represent the Q-statistics for the Ljung-Box test on the innovations w_k in the main rows with p-values in parentheses immediately below each main row. Similarly, the GLB labeled columns show the Q-statistics for the Ljung-Box test on the standardized residual e_k from fitting returns to a GARCH(1,1) model. Significance levels at 99% and 95% are signaled by two and one stars next to the main row statistics respectively. Under the random walk null hypothesis, the value of the variance ratio is 1, the Dickey-Fuller regression first lag coefficient is 1 and w_t's are independent. Under the random walk with GARCH residuals price hypothesis, the Dickey-Fuller regression first lag coefficient is 1, LB Q-statistics are significant and GLB Q-statistics are insignificant.

Category	Maturity	VR(2)	VR(5)	VR(10)	VR(20)	DF	LB(5)	LB(10)	GLB(5)	GLB(10)
Treasury	1-month	1.10*	0.96	1.00	1.04	1.00	79.68**	131.76**	59.42**	92.21**
Rate		(1.97)	(-0.41)	(0.01)	(0.17)	(0.84)	(0.00)	(0.00)	(0.00)	(0.00)
	3-month	1.03	0.96	1.08	1.22	1.00	93.48**	158.91**	94.55**	156.05**
		(0.74)	(-0.48)	(0.67)	(1.28)	(0.82)	(0.00)	(0.00)	(0.00)	(0.00)
	6-month	1.03	1.05	1.19	1.32*	1.00	36.31**	57.41**	56.27**	107.48**
		(0.82)	(0.66)	(1.67)	(2.02)	(0.84)	(0.00)	(0.00)	(0.00)	(0.00)
	1-year	0.99	0.95	0.97	0.95	1.00	16.34**	26.85**	11.41*	21.91*
		(-0.36)	(-0.65)	(-0.31)	(-0.28)	(0.78)	(0.01)	(0.00)	(0.04)	(0.02)
	2-year	1.01	0.93	0.92	0.89	1.00	11.65*	28.23**	10.76	18.27
		(0.22)	(-0.98)	(-0.75)	(-0.68)	(0.65)	(0.04)	(0.00)	(0.06)	(0.05)
	3-year	1.04	0.98	0.99	1.02	1.00	11.97*	26.93**	13.65*	19.16*
		(1.18)	(-0.35)	(-0.08)	(0.12)	(0.60)	(0.04)	(0.00)	(0.02)	(0.04)
	5-year	1.07*	1.03	1.04	1.07	1.00	18.38**	27.15**	15.20**	18.66*
		(2.55)	(0.51)	(0.46)	(0.56)	(0.53)	(0.00)	(0.00)	(0.01)	(0.04)
	7-year	1.06*	1.02	1.01	1.03	1.00	17.85**	24.08**	17.61**	20.00*
		(2.49)	(0.29)	(0.07)	(0.26)	(0.51)	(0.00)	(0.01)	(0.00)	(0.03)
	10-year	1.06**	1.02	1.01	1.04	1.00	17.79**	24.28**	14.59*	17.85
		(2.66)	(0.43)	(0.09)	(0.32)	(0.46)	(0.00)	(0.01)	(0.01)	(0.06)
	20-year	1.03	1.00	0.97	0.96	1.00	6.47	10.70	7.50	9.72
		(1.43)	(-0.06)	(-0.47)	(-0.32)	(0.53)	(0.26)	(0.38)	(0.19)	(0.47)
	30-year	1.06	1.05	0.99	0.95	1.00	8.44	10.30	7.37	10.21
		(1.76)	(0.78)	(-0.13)	(-0.33)	(0.48)	(0.13)	(0.41)	(0.19)	(0.42)

Table 1.16: The test statistics of Treasury debt interest rates for the period Jan 1, 1996 - Dec 31, 2005 for various maturities. The variance ratios $1 + M(q)$, $q = 2, 5, 10, 20$, are reported in VR labeled columns and in the main rows in line with the ETF tickers, whereas the corresponding z-scores are shown in parentheses immediately below each main row. The DF labeled column shows the first lag regression coefficients of log prices for the Dickey-Fuller test in the main rows with the corresponding p-values in parentheses immediately below the main rows. The LB labeled columns represent the Q-statistics for the Ljung-Box test on the innovations w_k in the main rows with p-values in parentheses immediately below each main row. Similarly, the GLB labeled columns show the Q-statistics for the Ljung-Box test on the standardized residual e_k from fitting returns to a GARCH(1,1) model. Significance levels at 99% and 95% are signaled by two and one stars next to the main row statistics respectively. Under the random walk null hypothesis, the value of the variance ratio is 1, the Dickey-Fuller regression first lag coefficient is 1 and w_t's are independent. Under the random walk with GARCH residuals price hypothesis, the Dickey-Fuller regression first lag coefficient is 1, LB Q-statistics are significant and GLB Q-statistics are insignificant.

sures have led to the mean-reverting behaviors in the Treasury debt market.

In Table 1.16, we summarize the test statistics for the same set of interest rates but for an earlier period January 1996 - December 2005. In this decade, the variance ratios are mostly greater than one. In several cases including 1-month, 6-month, 5-year, 7-year and 10-year, Treasury debts exhibit statistically significant positive autocorrelation. The Dickey-Fuller tests display no significant statistics; their p-values, however, show a decreasing pattern as the maturity of the Treasuries increase, which is the opposite to the results in Table 1.16. Rejections of the random walk hypothesis largely come from the Ljung-Box test. The GARCH conditional variance show little power in explaining the autocorrelations in the log-returns of the rates, leaving many statistically significant Q-statistics in the GLB labeled columns. The return-generating process seems complex and the source of inefficiency is not identified successfully.

Overall, the interest rate dynamics appear to be strikingly different from the period January 2006-September 2013 considered before. It is not clear if the current strong negative autocorrelation or mean-reverting behavior in the short-term rates will persist going forward.

1.3.2.2 *Bond ETFs and their tracking indices*

Among the bond ETFs we have considered, SHY, IEI, IEF, TLT, LQD, HYG and MBB are index ETFs. We compare the autocorrelation of the daily returns of these ETFs with those of their tracking bond indices for the period 2008-2013. The autocorrelation statistics and ETFs' tracking errors are reported in Table 1.17.

Bond ETFs' tracking errors range from 5 basis points (SHY) to 98 basis points (HYG). HYG having a higher tracking error may be due to its higher expense ratio (0.5%) and the fact that it takes a slightly more active approach in fund management. We also see that the autocorrelations of bond indices tend to be higher than those of the ETFs. In particular, the autocorrelation difference is rather small for Treasury debt based ETFs between 0.01 and 0.04, whereas for corporate bond based ETFs, this gap is quite sizable between 0.22 and 0.61. We will explore this feature in the

Ticker	Index	ETF	Index-ETF	Tracking Error
SHY	-0.08	-0.12	0.037	0.0005
IEI	-0.05	-0.06	0.013	0.0009
IEF	-0.03	-0.05	0.023	0.0016
TLT	-0.03	-0.04	0.014	0.0035
LQD	0.15	-0.07	0.222	0.0062
HYG	0.56	-0.05	0.608	0.0098
MBB	0.03	-0.06	0.092	0.0015

Table 1.17: The autocorrelations of bond ETFs with those of the ETFs' tracking indices for the period 2008-2013 are compared. The"Index" column contains the first lag autocorrelation of ETF's tracking index daily return. The next column labeled by "ETF" reports the corresponding ETF's autocorrelation using its daily returns. The"Index-ETF" column shows the difference between the index autocorrelation and the ETF autocorrelation. The last column shows the ETF's tracking error.

next subsection.

1.3.2.3 *Corporate bond and mortgage ETFs*

LQD and HYG are ETFs based on investment grade and high yield corporate bonds respectively. They both have statistically significant Ljung-Box Q-statistics, rejecting the random walk hypothesis. However, their variance ratios and Dickey-Fuller t-ratios fail to do so. In particular, HYG shows more random walk type behavior; its variance ratios are fairly close to 1, whereas LQD's variance ratios are much more smaller. Nonetheless, this seems to contradict the empirical results by Hong, Lin and Wu (2012) on corporate bond indices; they find significant positive autocorrelations of returns for both high yield and investment grade types. By applying our testing methods on the two ETFs' underlying indices, iBoxx Dollar Liquid Investment Grade Index and iBoxx Dollar Liquid High Yield Index, we are able to uncover the positive autocorrelation in this market shown in Table 1.18. Both indices display highly significant and large variance ratios at all lags, with HYG the most prominent. All Ljung-Box statistics reject their null hypothesis, as well.

We attempt to explain this strong positive autocorrelation of bond index by the illiquidity of the corporate bonds market. (1) Tucker [Tucker and Laipply (2013)] documented that on each business day, less than 10%

Index	VR(2)	VR(5)	VR(10)	VR(20)	DF	LB(5)	LB(10)	GLB(5)	GLB(10)
iBoxx Dollar Liquid	1.15*	1.35**	1.43*	1.82**	1.00	59.08**	96.82**	96.22**	106.50**
Investment Grade Index	(2.10)	(2.67)	(2.48)	(3.54)	(0.87)	(0.00)	(0.00)	(0.00)	(0.00)
iBoxx Dollar Liquid	1.56**	2.68**	3.65**	4.60**	1.00	859.22**	966.59**	427.00**	443.78**
High Yield Index	(7.49)	(10.91)	(12.22)	(12.35)	(0.98)	(0.00)	(0.00)	(0.00)	(0.00)
Barclays Capital	1.03	1.03	0.78	0.84	1.00	41.70**	54.03**	9.13	14.64
US MBS Index	(0.78)	(0.33)	(-1.60)	(-0.81)	(0.35)	(0.00)	(0.00)	(0.10)	(0.15)

Table 1.18: The test statistics of the underlying indices of LQD, HYG and MBB for the period Jan 1, 2008 - Sep 30, 2013. The variance ratios $1 + M(q)$, $q = 2, 5, 10, 20$, are reported in VR labeled columns and in the main rows in line with the ETF tickers, whereas the corresponding z-scores are shown in parentheses immediately below each main row. The DF labeled column shows the first lag regression coefficients of log prices for the Dickey-Fuller test in the main rows with the corresponding p-values in parentheses immediately below the main rows. The LB labeled columns represent the Q-statistics for the Ljung-Box test on the innovations w_k in the main rows with p-values in parentheses immediately below each main row. Similarly, the GLB labeled columns show the Q-statistics for the Ljung-Box test on the standardized residual e_k from fitting returns to a GARCH(1,1) model. Significance levels at 99% and 95% are signaled by two and one stars next to the main row statistics respectively. Under the random walk null hypothesis, the value of the variance ratio is 1, the Dickey-Fuller regression first lag coefficient is 1 and w_t's are independent. Under the random walk with GARCH residuals price hypothesis, the Dickey-Fuller regression first lag coefficient is 1, LB Q-statistics are significant and GLB Q-statistics are insignificant.

of the pool of underlying bonds held by HYG is actively traded. (2) When the underlying high yield bonds are not traded, their prices are estimated for the computation of the index NAV. These estimates might be based on similar securities' execution prices, levels of various interest rates and credit spreads. (3) The bid/ask spread for the period 9/30/2009 - 9/30/2012 is on average as high as 190 bps [Tucker and Laipply (2012)]. These observations may imply that liquidity factor can play an important role in the corporate bonds ETF markets. The non-synchronous model [Lo and MacKinlay (1988)] suggests the first two points lead to positive autocorrelation, whereas the third point reduces index autocorrelation from the bid/ask bounce argument [Roll (1984)]. Since the index consists of a basket of bonds, the positive autocorrelation contribution dominates via positive cross-correlations.

Similar analysis has been done for MBB, a mortgage based ETF. It shows some level of discrepancies, too. For instance, the variance ratios of its underlying index are all greater than 1 at lag 2 and 5, but those of MBB are less than 1. This suggests that the mortgage market can also be inefficient and that the mortgage ETFs may allow relevant information to be reflected in their prices faster than in the over-the-counter mortgage product trading. In addition, the difference between the variance ratios at lag 2 and lag 20 is quite significant for both MBB and its underlying index, which may indicate long-term mean reversion.

Since ETFs trade like stocks, investors can react rather quickly to new information in the credit and mortgage markets by trading the corresponding ETF products. As a result, overreactions may happen more often in the ETFs, showing negative serial correlations.

One interesting question derived from this inefficiency is : will the ETFs lead the corresponding credit and mortgage markets? From an operational perspective, this is very likely to be true. Although these ETFs are essentially a basket of bonds, they have exposure to both exchange as well as OTC markets. For instance, when a sell-off of corporate bonds takes place due to market developments such as hiking rates, unlike other types of pooled investments of corporate bonds, bond ETFs may first re-price and realize discounts relative to their NAV, rather than being forced to redemption immediately. During the recent crisis, the fund HYG experienced

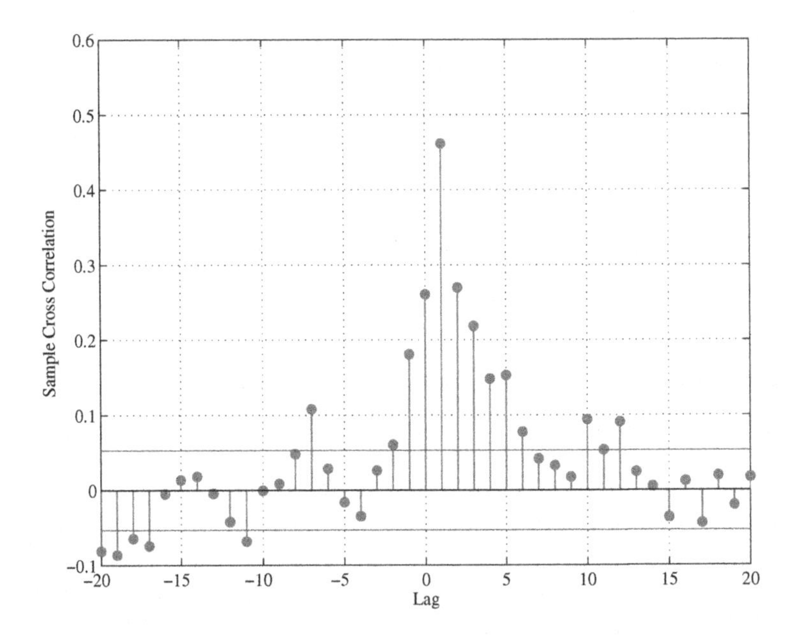

Fig. 1.6: The sample cross correlations between HYG daily returns and its underlying iBoxx Cash Liquid High Yield Index for the period January 2008 - September 2013.

similar dislocations between its price and its NAV, whereas Treasury debts based ETFs with more underlying market depth did not.

As a result, ETF prices on the exchange may potential lead those of the bond indices. To test this statistically, we evaluate the cross correlations between the returns of the ETFs and their tracking indices. HYG shows the most promise. Figure 1.6 plots the cross correlation values for lags from -20 to 20 for HYG. It shows significant sample cross correlation for lags from 0 to 5. This implies that HYG indeed leads its underlying tracking index up to 5 days. We also plot the cross correlation for LQD in Figure 1.7, but it only shows one lag significance. Judging from the underlying securities of HYG and LQD, we regard these observations reasonable, because the investment grade bonds are more liquidly traded compared with the high yield (junk) bonds. Similar analysis is done on MBB, as well. We do not yet see any evidence of lead-lag effects as in the credit ETFs. Nonetheless, it is worth noting that the indices tracked by these credit ETFs are not actually tradable.

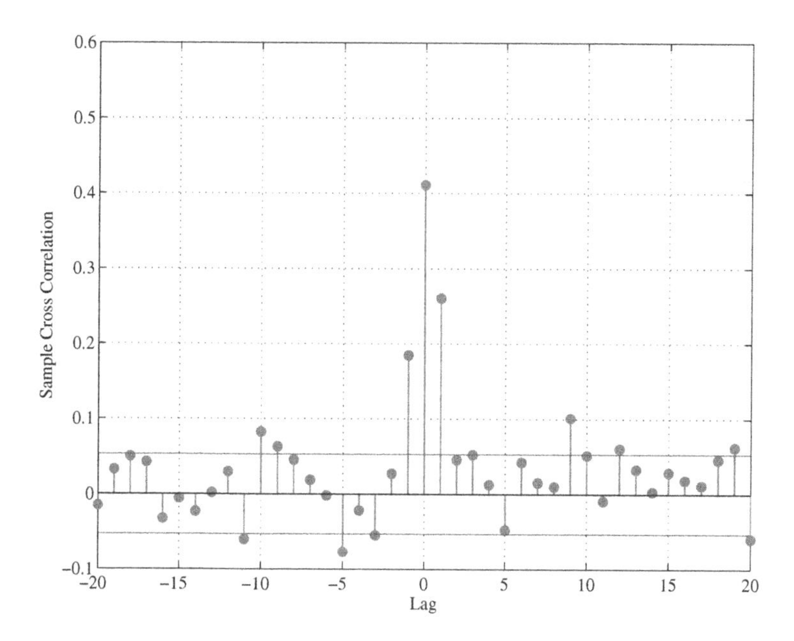

Fig. 1.7: The sample cross correlations between LQD daily returns and its underlying iBoxx Cash Liquid Investment Grade Index for the period January 2008 - September 2013.

To further explore the potential lead-lag effects. We resort to high yield bond mutual funds. We have collected 7 mutual funds whose historical data date back to 2008. The tickers and the names of the 7 mutual funds are listed in Table 1.19. They are within the top 10 best mutual funds ranked by US News among a list of 174 high yield mutual fund. The ranking is based on leading fund industry researchers. These mutual funds offer investors to similar underlying bond markets as HYG. For illustration, Figure 1.8 shows the cumulative returns of HYG and USHYX in comparison. However, these mutual funds are only available to trade once a day after market close. We perform similar tests as before on their end of day prices obtained from Yahoo Finance and summarize the statistics in Table 1.20. We see that the table shows very strong test statistics with variance ratios all above 1, with only one of them (JIHDX at lag 2) not statistically significant. The significantly positive autocorrelations of these mutual fund returns are consistent with the high yield corporate bond index tracked by HYG.

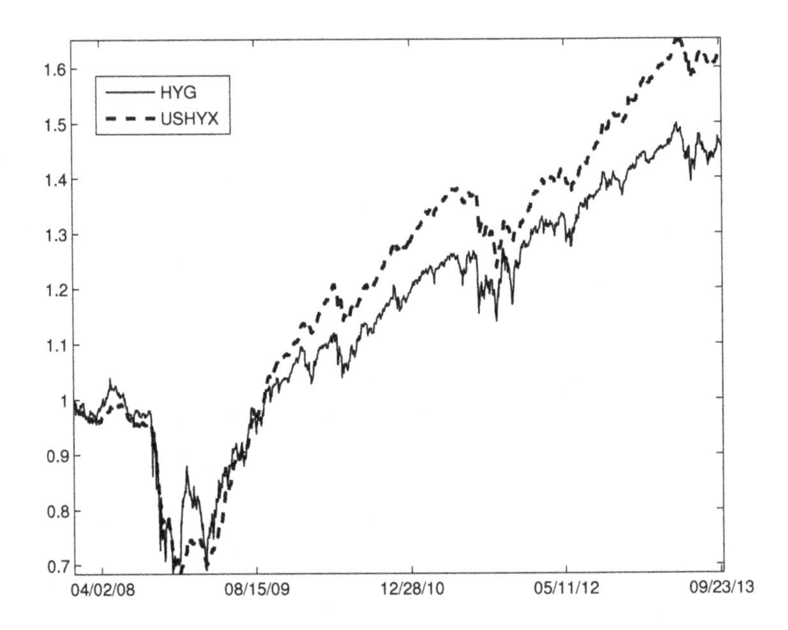

Fig. 1.8: The cumulative returns of HYG and USHYX for January 2008 - September 2013.

Ticker	Fund Name
LSHIX	Loomis Sayles Institutional High Income Fund
USHYX	USAA High Income Fund
UNHIX	Waddell & Reed Advisor High-Income Fund Class A
WHIAX	Ivy High Income Fund Class A
JIHDX	John Hancock II High Yield Fund Class 1
TRHYX	T. Rowe Price Institutional High Yield Fund
BHYAX	BlackRock High Yield Bond Portfolio Investor A Shares

Table 1.19: The tickers and names of the 7 high yield bond mutual funds.

Moreover, HYG leads the performance of these mutual funds as it did to its underlying index. We plot the cross-correlation plot in Figure 1.9. This plot represents the general relationship between HYG and all 7 mutual funds listed here. The negative lags typically do not exhibit statistical significance with sample cross correlations lying within the two horizontal

Ticker	VR(2)	VR(5)	VR(10)	VR(20)	DF	LB(5)	LB(10)	GLB(5)	GLB(10)
LSHIX	1.34**	2.03**	2.84**	3.89**	1.00	355.75**	451.38**	190.21**	230.35**
	(5.45)	(7.52)	(9.17)	(10.72)	(0.99)	(0.00)	(0.00)	(0.00)	(0.00)
USHYX	1.50**	2.66**	4.11**	6.10**	1.00	1000.99**	1308.90**	386.50**	467.71**
	(8.68)	(13.28)	(16.56)	(19.80)	(1.00)	(0.00)	(0.00)	(0.00)	(0.00)
UNHIX	1.38**	2.23**	3.18**	4.28**	1.00	528.04**	636.76**	172.57**	195.40**
	(6.24)	(9.17)	(10.92)	(12.26)	(1.00)	(0.00)	(0.00)	(0.00)	(0.00)
WHIAX	1.38**	2.22**	3.20**	4.25**	1.00	532.11**	656.00**	164.12**	195.87**
	(5.34)	(7.62)	(9.22)	(10.28)	(1.00)	(0.00)	(0.00)	(0.00)	(0.00)
JIHDX	1.09	1.51*	2.08**	2.84**	1.00	110.38**	213.28**	316.22**	349.18**
	(0.67)	(2.15)	(3.64)	(4.89)	(0.98)	(0.00)	(0.00)	(0.00)	(0.00)
TRHYX	1.48**	2.42**	3.33**	4.37**	1.00	648.49**	742.19**	318.44**	342.13**
	(8.79)	(12.45)	(13.75)	(14.60)	(0.99)	(0.00)	(0.00)	(0.00)	(0.00)
BHYAX	1.39**	2.21**	3.05**	4.03**	1.00	481.42**	565.28**	250.45**	279.43**
	(6.06)	(8.41)	(9.62)	(10.71)	(0.99)	(0.00)	(0.00)	(0.00)	(0.00)

Table 1.20: The test statistics of 7 high yield bond mutual funds for the period Jan 1, 2008 - Sep 30, 2013. The variance ratios $1 + M(q)$, $q = 2, 5, 10, 20$, are reported in VR labeled columns and in the main rows in line with the ETF tickers, whereas the corresponding z-scores are shown in parentheses immediately below each main row. The DF labeled column shows the first lag regression coefficients of log prices for the Dickey-Fuller test in the main rows with the corresponding p-values in parentheses immediately below the main rows. The LB labeled columns represent the Q-statistics for the Ljung-Box test on the innovations w_k in the main rows with p-values in parentheses immediately below each main row. Similarly, the GLB labeled columns show the Q-statistics for the Ljung-Box test on the standardized residual e_k from fitting returns to a GARCH(1,1) model. Significance levels at 99% and 95% are signaled by two and one stars next to the main row statistics respectively. Under the random walk null hypothesis, the value of the variance ratio is 1, the Dickey-Fuller regression first lag coefficient is 1 and w_t's are independent. Under the random walk with GARCH residuals price hypothesis, the Dickey-Fuller regression first lag coefficient is 1, LB Q-statistics are significant and GLB Q-statistics are insignificant.

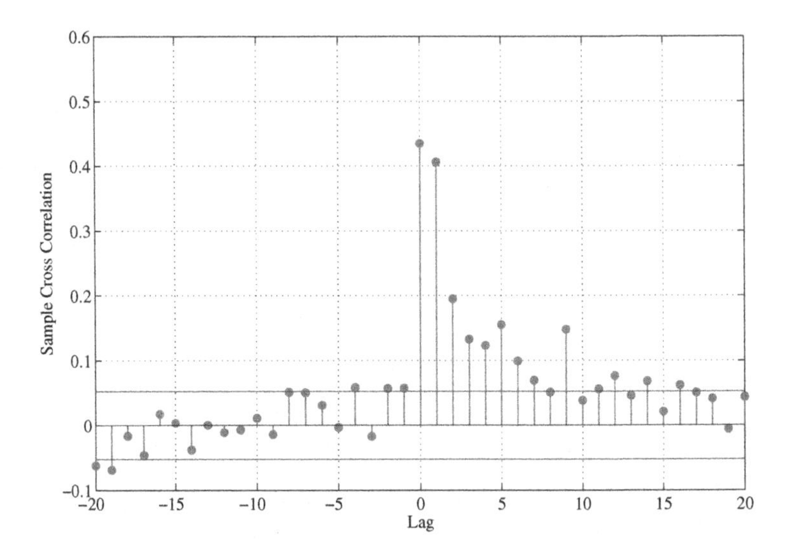

Fig. 1.9: The sample cross correlations between HYG daily returns and the mutual fund USHYX daily returns for the period January 2008 - September 2013.

lines which represent the 5% confidence bands. However, the positive lags show significant cross correlation, suggesting the returns of HYG lead the returns of those mutual funds. The peak of sample correlations take place at lag 0, meaning HYG and the mutual funds are highly correlated on a daily basis. This lead-lag effect also suggests that ETF allows information on high yield bonds to be more rapidly absorbed by the investors.

1.3.3 *Commodity ETFs*

Commodities are a separate asset class from stocks and bonds, which offers additional diversification for investors in the financial market. The idea of a gold exchange traded fund was first proposed by Benchmark Asset Management Company Private Ltd[18] in India in 2002 but was launched rather later in 2007. The world's first commodity ETFs was actually the

[18]As of July 14, 2011, Benchmark Asset Management Company Private Limited (BAMC) is a part of the Goldman Sachs group. BAMC is an asset management company in India with a primary focus on indexing and using quantitative methods to create innovative financial products.

Gold Bullion Securities listed on the Australian stock exchange in 2003. Commodity ETFs in the US were created slightly afterwards.

The commodity ETFs often invest in commodities such as precious metals, agricultural products or natural resources. Since most of the commodity ETFs do not invest in securities, commodity ETFs are typically not regulated as investment companies under the Investment Company Act of 1940 in the US but by the Commodity Futures Trading Commission (CFTC) under the Commodity Exchange Act and by the SEC under the Securities Act of 1933.

For different targeting assets, the ETFs in commodities are created differently. For instance, some commodity ETFs are structured by holding the actual physical underlying assets such as GLD (SPDR Gold Trust), some are composed as a synthetic which holds the corresponding commodity futures such as USO (United states Oil Fund) as they are less convenient to store, others are a hybrid of both physical assets and derivative positions in their underlying commodities. Apart from the ETFs which invest in only one commodity, there are also those which track a specific commodity index like GSG (iShares GSCI Commodity-indexed Trust).

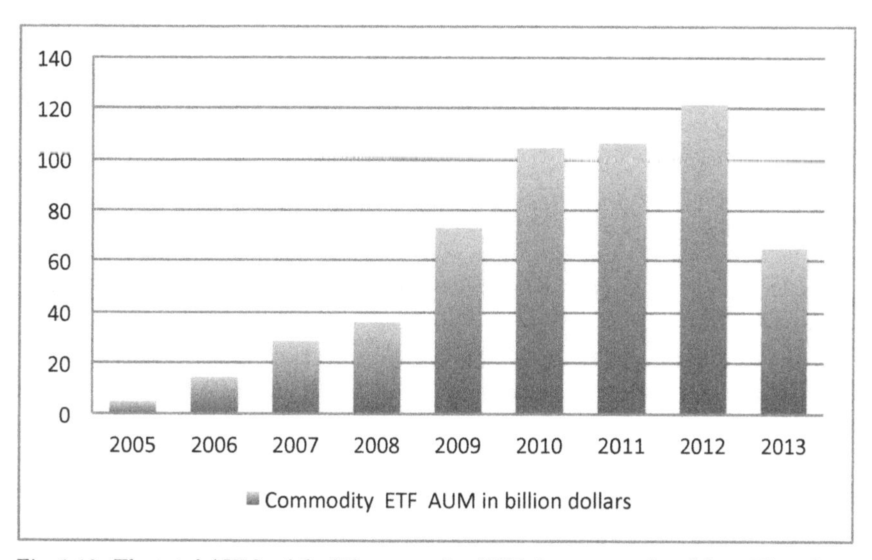

Fig. 1.10: The total AUMs of the US commodity ETFs by year retrieved from Bloomberg for 2005-2013.

Commodity ETFs are very well-received in the markets, as well. Getting exposure to commodities can often be difficult due to their accessibility and storage costs. To trade commodity futures, investors also need to understand their delivery mechanism. Commodity ETFs can offer quite a lot of convenience. Take the gold ETF GLD as an example. Since the fund

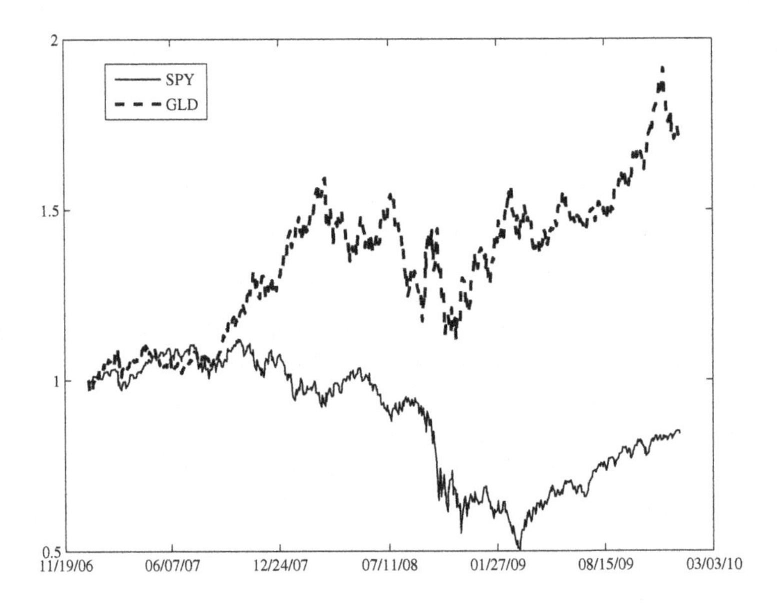

Fig. 1.11: The cumulative returns of the gold ETF, GLD, and the SPDR S&P 500 Index ETF, SPY, during the recent financial crisis, are compared.

is backed by physical gold, holding shares of GLD is equivalent to holding gold adjusted to its market value. In the sub-prime mortgage financial crisis, this ETF took off as a solid safe haven with a great growth of AUM. In August 2011, GLD had over $77 billion asset under management, even surpassing one of the largest ETFs, SPY, for a relatively short period. Figure 1.11 compares the performances of GLD and SPY for the period of crisis and early recovery. However, as the equity market and the US economy generally recovered in 2013, money flew out of GLD. Figure 1.10 displays the AUMs of US commodity ETFs. We see that the commodity ETFs have achieved rather steady growth since 2005, except for the year 2013. In 2013, the AUM of the total US commodity ETFs dropped from $120 billion to just above $61 billion. In fact, over 99% of this money outflow took place in GLD.

In this section, we investigate six commodity ETFs, covering gold, silver, oil, gas and agriculture. Table 1.21 reports the test results for the period Jan 1, 2008 - Sep 30, 2013 for these ETFs. Overall, it is hard to reject the random walk null hypothesis for this asset class. Only two ETFs, UNG and DBA, show significant Ljung-Box Q-statistics rejecting the independence in log-returns. However, after fitting their conditional variances to GARCH, the residuals do not exhibit autocorrelations any more as seen from their Q-statistics in the GLB columns. UNG is shown to be the only case where a rejection of the variance ratio is found and it is at lag 2. Furthermore, the variance ratios are no longer less than 1; quite a number of them are equal or exceeding 1. The monotonicity of variance ratios in q is not typical as in equity ETFs. GSP even shows an increasing pattern of variance ratios in q reaching 1.12 at $q = 20$, which is rather different from what we have seen for equity ETFs for the same period of time. The Dickey-Fuller test, however, remains powerless.

The reported collection of 6 commodity ETFs is only a small sample of the commodity universe. We have not covered more ETFs due to the fact that most of the commodity ETFs are created rather recently. Among our collection, GLD and SLV represent the class of precious metals. They are both physical funds and track the spot prices of the two commodities. We have found that the log price of the gold and silver spots behave like random walks for the past five years. Smith (2002) obtained similar results for the London gold daily closing prices for the period Jan 3, 1990 - Sep 27, 2001.

USO and UNG are ETFs which offer investors the exposure to crude oil and natural gas markets managed by United States Commodity Funds LLC[19]. Unlike the two precious metal ETFs we have seen earlier, these two energy ETFs are created synthetically using commodity futures contracts, as the underlying commodities are harder to store. The front month futures

[19]United States Commodity Funds LLC (USCF) is a US company which specializes in managing commodity ETFs. USCF was one of the earliest issuers of exchange-traded commodity funds in the US. Currently, the ETFs managed by USCF include USO, USL, DNO, BNO, UNG, UNGL, UGA, UHN, USCI, USAG, USMI and CPER.

Category	Ticker	VR(2)	VR(5)	VR(10)	VR(20)	DF	LB(5)	LB(10)	GLB(5)	GLB(10)
Commodity	GLD	1.01	0.98	0.94	0.87	1.00	2.50	7.50	4.08	15.90
		(0.33)	(-0.21)	(-0.50)	(-0.76)	(0.51)	(0.78)	(0.68)	(0.54)	(0.10)
	SLV	1.01	0.99	1.01	1.00	1.00	0.87	1.79	6.02	13.10
		(0.16)	(-0.08)	(0.05)	(-0.01)	(0.56)	(0.97)	(1.00)	(0.30)	(0.22)
	USO	0.96	0.96	0.94	1.06	1.00	6.92	8.34	3.46	4.21
		(-1.08)	(-0.53)	(-0.50)	(0.29)	(0.29)	(0.23)	(0.60)	(0.63)	(0.94)
	UNG	0.93*	0.95	0.92	0.90	1.00	18.68**	22.04*	7.02	8.37
		(-2.42)	(-0.71)	(-0.78)	(-0.67)	(0.66)	(0.00)	(0.01)	(0.22)	(0.59)
	DBA	0.95	0.88	0.86	0.91	0.99	13.50*	19.27*	9.16	10.70
		(-1.63)	(-1.50)	(-1.06)	(-0.45)	(0.31)	(0.02)	(0.04)	(0.10)	(0.38)
	GSP	0.97	1.00	1.01	1.12	1.00	8.20	9.88	0.68	2.71
		(-0.99)	(-0.06)	(0.05)	(0.69)	(0.32)	(0.15)	(0.45)	(0.98)	(0.99)

Table 1.21: The test statistics of 6 commodity ETFs for the period Jan 1, 2008 - Sep 30, 2013. The variance ratios $1 + M(q)$, $q = 2, 5, 10, 20$, are reported in VR labeled columns and in the main rows in line with the ETF tickers, whereas the corresponding z-scores are shown in parentheses immediately below each main row. The DF labeled column shows the first lag regression coefficients of log prices for the Dickey-Fuller test in the main rows with the corresponding p-values in parentheses immediately below the main rows. The LB labeled columns represent the Q-statistics for the Ljung-Box test on the innovations w_k in the main rows with p-values in parentheses immediately below each main row. Similarly, the GLB labeled columns show the Q-statistics for the Ljung-Box test on the standardized residual e_k from fitting returns to a GARCH(1,1) model. Significance levels at 99% and 95% are signaled by two and one stars next to the main row statistics respectively. Under the random walk null hypothesis, the value of the variance ratio is 1, the Dickey-Fuller regression first lag coefficient is 1 and w_t's are independent. Under the random walk with GARCH residuals price hypothesis, the Dickey-Fuller regression first lag coefficient is 1, LB Q-statistics are significant and GLB Q-statistics are insignificant.

contracts of crude oil and natural gas tend to track their spot levels[20] quite well (see Figure 1.12 for natural gas), however, futures contracts expire. When this occurs, the exchange traded fund managers have to liquidate the existing commodity futures position and use the cash to purchase the next available futures contract.[21]

When the front month contracts are gradually rolled into the next month contracts, the exchange traded fund may experience a dislocation in tracking the underlying asset prices. In particular, this dislocation can be in the form of either a roll cost or a roll yield depending on the shape of the futures term structure. If the term structure is in contango, then a cost is realized which reduces the funds NAV and as a result the market price will fall. If the term structure is in backwardation, then a roll yield is achieved. Figure 1.12 compares the cumulative returns of the natural gas spot, front month futures and the exchange traded fund UNG. We see that UNG considerably underperform the tracking spot for the period 2009-2013.

We also perform the variance ratio tests again on the corresponding spot price indices in comparison with the ETFs' statistics. The results are reported in Table 1.22. From the table, we see little difference between the crude oil index and its ETF, whereas, the inconsistency in natural gas is rather significant. The spot returns of natural gas tend to have small positive autocorrelation, whereas those of the ETF showed significant negative autocorrelation. The gap between the first lag autocorrelation of these two is as wide as 0.13 for the period 2008-2013 in aggregate. In addition, the GLB columns for the natural gas spot index show strongly significant Ljung-Box Q-statistics, different from the UNG ETF's behavior.

We attempt to explain this autocorrelation gap between the natural gas spot and UNG by the following model. Denote the spot prices by S_t, the prices of the front month futures by F_t, the back month futures by B_t at time t, and the weight of the back month futures in the rolling mechanism of generating the level of these ETFs by w_{t-1} . Then we have

$$F_t = S_t e^{c_t \tau} \tag{1.44}$$

[20] They are the WTI Crude Oil Spot Price and the Henry Hub Gulf Coast Natural Gas Spot Price respectively for USO and UNG.

[21] For these two funds in particular, the rolling takes place monthly roughly two weeks before the expiration of the front month futures contract. The roll-over lasts for about four days.

Spot Index	VR(2)	VR(5)	VR(10)	VR(20)	DF	LB(5)	LB(10)	GLB(5)	GLB(10)
Crude Oil	1.00	1.04	0.91	0.94	0.99	28.48**	42.64**	1.35	2.92
	(0.09)	(0.39)	(-0.59)	(-0.26)	(0.33)	(0.00)	(0.00)	(0.93)	(0.98)
Natural Gas	1.06	0.92	0.86	0.80	1.00	28.82**	44.71**	20.29**	24.92**
	(1.02)	(-0.71)	(-0.80)	(-0.85)	(0.27)	(0.00)	(0.00)	(0.00)	(0.01)

Table 1.22: The statistics of the crude oil and natural gas spot indices for the period Jan 1, 2008 - Sep 30, 2013. The variance ratios $1 + M(q)$, $q = 2, 5, 10, 20$, are reported in VR labeled columns and in the main rows in line with the commodity names, whereas the corresponding z-scores are shown in parentheses immediately below each main row. The DF labeled column shows the first lag regression coefficients of log prices for the Dickey-Fuller test in the main rows with the corresponding p-values in parentheses immediately below the main rows. The LB labeled columns represent the Q-statistics for the Ljung-Box test on the innovations w_k in the main rows with p-values in parentheses immediately below each main row. Similarly, the GLB labeled columns show the Q-statistics for the Ljung-Box test on the standardized residual e_k from fitting returns to a GARCH(1,1) model. Significance levels at 99% and 95% are signaled by two and one stars next to the main row statistics respectively. Under the random walk null hypothesis, the value of the variance ratio is 1, the Dickey-Fuller regression first lag coefficient is 1 and w_t's are independent. Under the random walk with GARCH residuals price hypothesis, the Dickey-Fuller regression first lag coefficient is 1, LB Q-statistics are significant and GLB Q-statistics are insignificant.

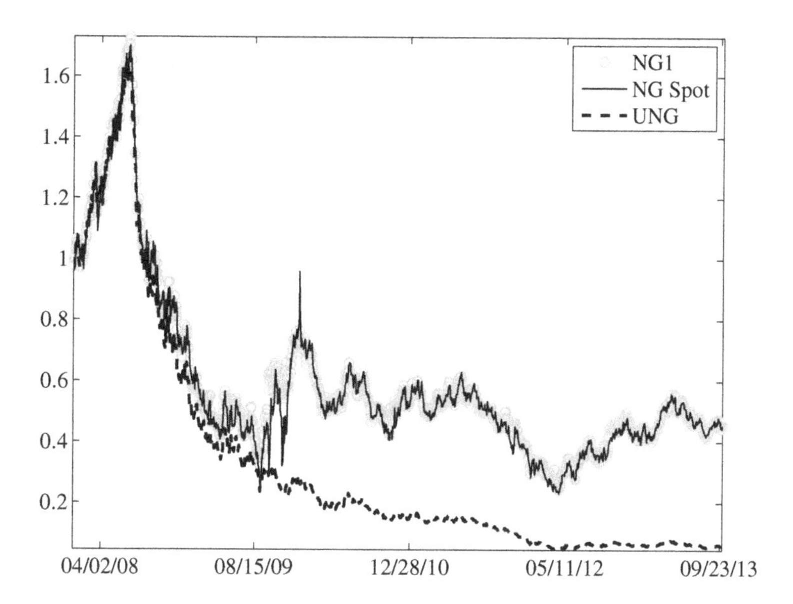

Fig. 1.12: The cumulative returns of UNG, the natural gas spot and the front month futures contracts are compared for the period 2008-2013. UNG is represented by the dashed black line. The first month futures contracts are represented by circles in the grey color. The natural gas spots are represented by the solid black line.

and

$$B_t = S_t e^{c_t(\tau + \frac{1}{12})}, \tag{1.45}$$

where τ is the time to maturity of the front month futures and c_t is the annualized storage cost minus convenient yield at time t, assuming interest rates indistinguishable from zero.

Also, the price of the ETF, P_t, will follow

$$P_t = P_{t-1}\frac{(1 - w_{t-1})F_t + w_{t-1}B_t}{(1 - w_{t-1})F_{t-1} + w_{t-1}B_{t-1}}. \tag{1.46}$$

Prop 1.1. If we define the log return of P_t as p_t and that of S_t as r_t, then

$$p_t = \left(\frac{w_{t-1}}{1 + (s_{t-1} - 1)w_{t-1}} + \frac{12\tau}{s_{t-1}}\right)\Delta s_t + \frac{c_{t-1}}{252} + r_t + O\left(|\Delta s_t|^2\right),$$

where $s_t = \frac{B_t}{F_t}$ is the slope of the term structure of the futures.

commodity	ρ_p	ρ_r	e_r	η_u	η_{ur}	η_{ru}	$\rho_p - \rho_r$
crude oil	-0.043	0.001	0.001	-0.043	0.032	-0.033	-0.043
natural gas	-0.070	0.056	0.047	-0.788	0.723	-0.107	-0.125

Table 1.23: The components of the autocorrelation of USO and UNG returns for the period Jan 1, 2008 - Sep 30, 2013 based on the formulation of $p_t = r_t + u_t$. The autocorrelation can be represented as a summation of 5 terms as in $\rho_p = \rho_r + e_r + \eta_u + \eta_{ur} + \eta_{ru}$, where $\rho_r = \frac{Cov(r_t,r_{t-1})}{Var(r_t)}$, $e_r = \frac{Cov(r_t,r_{t-1})}{Var(p_t)} - \rho_r$, $\eta_u = \frac{Cov(u_t,u_{t-1})}{Var(p_t)}$, $\eta_{ru} = \frac{Cov(r_{t-1},u_t)}{Var(p_t)}$ and $\eta_{ur} = \frac{Cov(u_{t-1},r_t)}{Var(p_t)}$. The first three terms represent the influences of the autocorrelation of the spot returns and that of the storage cost changes. The last two terms represent the influences of the lead-lag relation between the change in storage costs and the spot returns.

The proof of this proposition is in the appendix of this chapter.

Now, it is easy to study the factor impacts on autocorrelation of p_t. For simplicity and tractability, we may write

$$p_t = r_t + u_t, \tag{1.47}$$

where u_t represents essentially the effects of the change in the slope of the futures term structure and the rolling mechanism. Then, the autocorrelation of p_t, denoted by ρ_p, can be expressed by the summation of five terms.

$$\rho_p = \rho_r + e_r + \eta_u + \eta_{ur} + \eta_{ru}, \tag{1.48}$$

where $\rho_r = \frac{Cov(r_t,r_{t-1})}{Var(r_t)}$, $e_r = \frac{Cov(r_t,r_{t-1})}{Var(p_t)} - \rho_r$, $\eta_u = \frac{Cov(u_t,u_{t-1})}{Var(p_t)}$, $\eta_{ru} = \frac{Cov(r_{t-1},u_t)}{Var(p_t)}$ and $\eta_{ur} = \frac{Cov(u_{t-1},r_t)}{Var(p_t)}$. The first three terms represent the influences of the autocorrelation of the spot returns and that of the storage cost changes. The last two terms represent the influences of the lead-lag relation between the change in storage costs and the spot returns.

Table 1.23 reports the estimates of these variables for crude oil and natural gas. We observe that the values for all variables of crude oil remain rather small in magnitude. For natural gas, however, we observe significant contribution of η_u and η_{ur}. This suggests that the storage cost and roll mechanism have strong mean reverting behavior and their changes lead the changes in the natural gas spot prices.[22] However, these two terms cancel

[22]Note that the lead-lag effect is consistent with Nick (2013), where he found that the natural gas futures European markets lead the spots, as Δc_t is the essential difference between futures returns and spot returns in our model.

each other in general. Note that the sum of η_u and η_{ur} is proportional to $Cov(u_{t-1}, p_t)$. This says that the term structure and the futures rolling operation do not cross-correlated much with the UNG's returns. The term η_{ru} is relatively strong and it makes up the major portion of the difference $\rho_p - \rho_r$. This implies that when natural gas spots increases, it tends to be associated with the fall of storage costs or the rise of convenience yield, flattening the natural gas futures' term structure.

Figure 1.13 plots the natural gas spot prices together with their front and back month futures prices for a specific period Sep 2012 - Dec 2012. It reflects the movements of the natural gas prices in response to weather

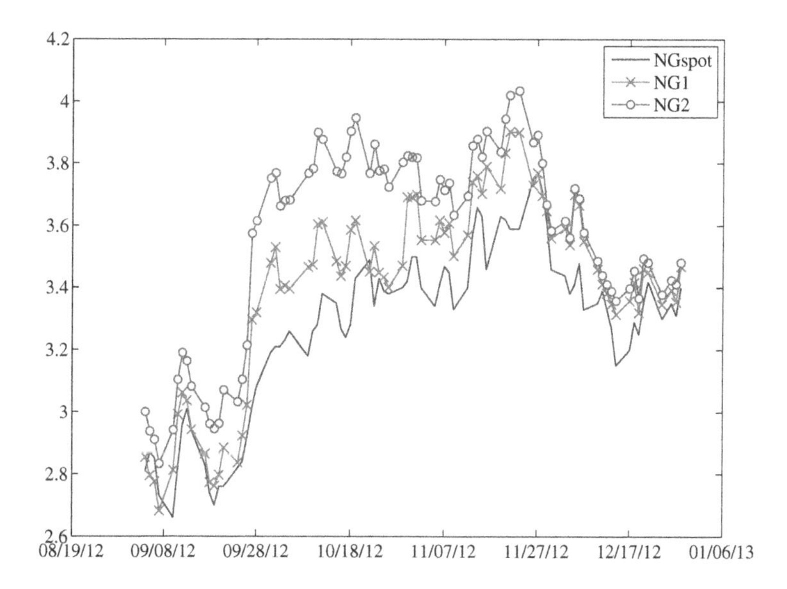

Fig. 1.13: The natural gas spot prices together with their front and back month futures prices for the period Sep 2012 - Dec 2012.

"forecasts calling for lower temperatures in the coming weeks" reported by the Wall Street Journal on Oct 1, 2012. It suggested a pickup in demand for gas-fired heating for the winter of 2012. In the figure, we see significant spikes in the natural gas futures prices in late September and early October in 2012. As suggested by our correlation statistics in Table 1.23, the jump in futures prices soon pull up the spot gas prices. As the spot prices rose throughout November, the gap between the natural gas futures and spot narrowed down, confirming the negativity of η_{ru}. The mean reversion of the futures' term structure is consistent with the sign of η_c.

Hence, it is natural to argue that if we were to see strong negative autocorrelation in the returns of UNG, we need the presence of big changes in natural gas futures term structure. However, due to the oversupply of natural gas in the marketplace, the most likely movement in the futures term structure has to be contango. According to United States Energy Information Administration, natural gas has seen to be significantly undervalued especially after the financial crisis. In Figure 1.14, we see that the dollars per million BTU[23] of crude oil has significantly surpassed those of natural gas since 2005. The source of the oversupply of natural gas tends

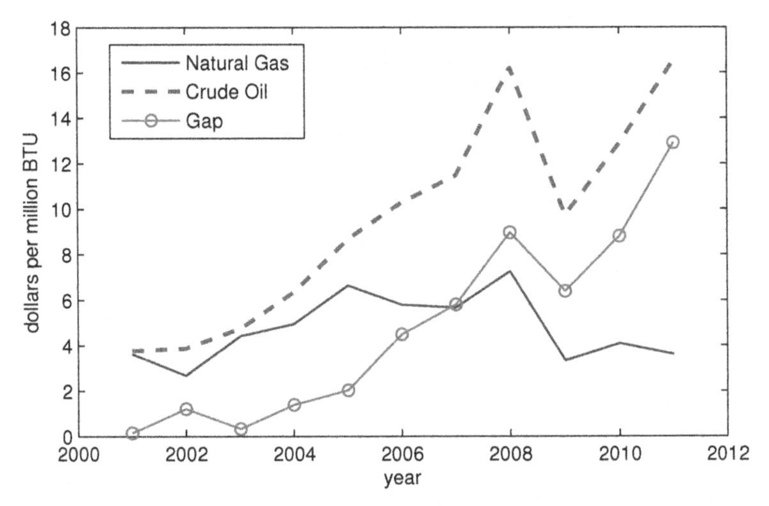

Fig. 1.14: The dollars per million BTUs of natural gas (solid line) and crude oil (dashed line) for the period 2000-2011. The difference between crude oil and natural gas is shown as the circled line.

to be associated with the discoveries of shale-related natural gas and the new technologies which allows its production. An article from CME Group [Azzarello (2014)] documents that shale-related natural gas production has increased by 417% between 2007 and 2012, which made up a significant portion of the total 20% increase in the natural gas production for the same

[23]BTU is short for the British thermal unit.

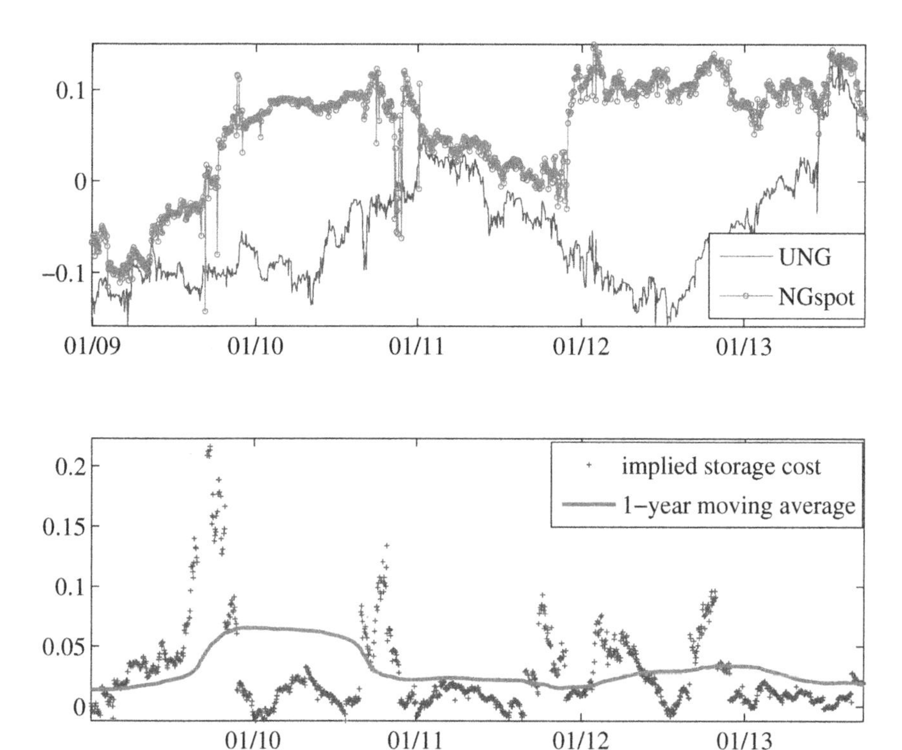

Fig. 1.15: The rolling first lag autocorrelations of the natural gas ETF and its spot with a window of 1-year in comparison with the corresponding implied storage cost (minus convenience yield). The upper panel displays the autocorrelations of UNG and natural gas spot returns for the period 2008-2013, whereas the lower panel shows the corresponding oil futures term structure implied storage cost (minus convenience yield) and its 1 year moving average.

period. The excess supply of natural gas however did not meet the desired demand. The 07-08 financial crisis led the US to recession and reduced the energy need in the expectation of a slow-growing economy. The natural gas market has not yet recovered to its pre-crisis level.

Figure 1.15 empirically confirms our prediction. The upper panel displays the autocorrelations of UNG and natural gas spot returns for the period 2008-2013, whereas the lower panel shows the corresponding oil futures term structure implied storage cost (minus convenience yield), i.e. $c_t/12$, and its 1-year moving average. By comparing both panels, we observe that the lower levels of the daily UNG return autocorrelations tend to take place when the 1-year moving average of the term-structure-implied storage costs are high. In addition, since the storage cost can also be interpreted as the slope of the futures term structure, we see that natural gas futures tend to be in contango rather frequently. Often, these contangos occur in winter.

DBA and GSP are ETFs based on commodity futures composites rather than individual products. DBA tracks the DBIQ Diversified Agriculture Index Excess Return consisting of some of the most liquid agricultural commodity futures, whereas GSP tracks the S&P GSCI Commodity Index covering all aspects of commodity futures. By reading the various test statistics, it is not evident that these index based commodity ETFs are different from the rest. We overall fail to reject the random walk model for commodity ETFs.

The efficiency of the energy commodity markets can be a challenging matter as it involves governments, large-scale producers, institutional investors and customers. Their supply and demand may be influenced by war, regional financial distress, international relations, etc. The existing literature on crude oil and natural gas prices in general have shown that the energy commodity markets are not uniformly efficient and are subject to time-varying autocorrelation for various time scales. For instance, Geman (2007) shows that the mean-reversion pattern prevails over the period 1994 - 2000 for both crude oil and natural gas but turns into a random walk during the period 2000 - 2004. Wang and Liu (2010) report that the WTI crude oil market short-term, medium-term and long-term behaviors are generally becoming efficient, however, the change of long-term correlations are more fierce than short-term ones, indicating that small fluctuations of this market can be forecasted better than large fluctuations. They conclude that short-term dynamics are mainly driven by the exogenous factors while demand and supply mechanism plays a main role in the long term. As a result, many asset managers tend to adopt the trend following strategies in trading commodities for alphas.

1.3.4 *Currency ETFs*

Currency ETFs were first created in the US market in 2005, offering investors the opportunity to trade global currencies on the exchange rather than using futures or other derivative securities. The pioneer fund of this kind is the Euro Currency Trust, FXE, launched in New York by Rydex Investments, offering the exposure to the cross rate of EURUSD. It has since become one of the most popular currency ETFs. Early currency ETFs concentrated on currencies in the developed world; emerging currency ETFs came later. Figure 1.16 plots the total AUMs of the currency ETFs in the US since the inception of currency ETFs. We see that the currency ETFs compared with the ETFs in other categories remain less developed. The AUM by the end of 2013 is only about $4.5 billion. A potential reason for this is that the expense ratios of these currency ETFs might be high relatively to interest rates.

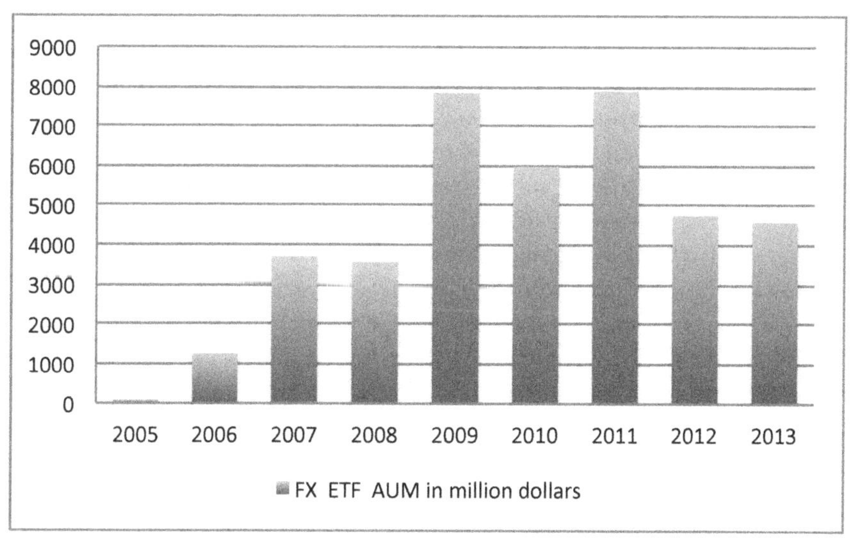

Fig. 1.16: The total AUMs of the currency ETFs in the US by year retrieved from Bloomberg for 2005-2013.

In our research, we will consider nine major currency ETFs including FXA (Australia), FXC (Canada), FXF (Swiss), FXE (EU), FXY (Japan),

FXB (UK), FXS (Sweden), CNY[24] (China) and INR (India), with presence of both currencies in developed markets and emerging markets. We perform the proposed hypothesis tests on these nine currency ETFs. Table 1.24 reports the corresponding statistics for the period January 2008 - September 2013.

We find that currency ETFs for the developed nations exhibit much weaker statistics in comparison with those from the emerging markets such as China and India. The variance ratios for developed economics are quite close to 1, whereas those of China and India are much lower. The z-scores for CNY and INR are significant even for $q = 20$ and monotonic decreasing as q level rises. The Dickey-Fuller tests for CNY and INR have strong confidence rejecting the unit root hypothesis and establishing stationarity. The Ljung-Box Q-statistics of developed currencies show certain level of rejections sporadically but after fitting the conditional variance to GARCH models, the residuals do not show further significant deviation from the random walk behavior. Yet, for CNY and INR the Q-statistics are strongly significant for both the LB and the GLB columns, suggesting complex return-generating dynamics.

The above observation may result from the fact that the exchange rates in China and India are subject to stronger governmental control and do not float the same way as the currencies of the developed world which adopts the free-floating currency model, thus contain more statistical information. Chiang et al. (2010) have similar results. They re-examine the validity of the weak form efficient market hypothesis for foreign exchange markets in four neighboring Asian economies including Japan, South Korea, Taiwan and Philippines. Their results show that the random walk patterns of the exchange rate return series cannot be rejected, except for Taiwan, where the free-floating rate is not adopted. Overall, currencies of the developing world are typically managed so that they are cheaper than those of the developed world. Such undervaluation in general offers advantage of the ex-

[24]CNY and INR are actually exchange-trade notes or short for ETNs, which are very similar to ETFs. Typically, both ETFs and ETNs are designed to track certain underlying indices. However, when you invest in an ETF, you are investing in a fund which holds the assets tracking or constituting the underlying index, such as stocks, bonds, commodities or their corresponding futures contracts, whereas an ETN is more like a bond. It is often an unsecured note issued by an institution. Therefore, ETNs are subject to default risks.

Category	Ticker	VR(2)	VR(5)	VR(10)	VR(20)	DF	LB(5)	LB(10)	GLB(5)	GLB(10)
Currency	FXA	0.94	0.91	0.87	0.89	1.00	14.91*	24.48**	6.48	9.13
		(-1.17)	(-0.77)	(-0.73)	(-0.43)	(0.62)	(0.01)	(0.01)	(0.26)	(0.52)
	FXC	1.00	1.07	1.01	0.86	0.99	14.03*	25.48**	5.41	14.86
		(0.03)	(0.83)	(0.11)	(-0.77)	(0.25)	(0.02)	(0.00)	(0.37)	(0.14)
	FXF	0.99	0.96	1.00	1.05	1.00	4.39	11.59	4.16	10.17
		(-0.26)	(-0.57)	(-0.03)	(0.32)	(0.37)	(0.49)	(0.31)	(0.53)	(0.43)
	FXE	1.03	1.02	1.07	1.13	0.99	3.36	15.18	1.96	6.20
		(0.95)	(0.34)	(0.63)	(0.82)	(0.17)	(0.64)	(0.13)	(0.85)	(0.80)
	FXY	0.97	0.89	0.84	0.84	1.00	7.59	14.91	2.62	11.33
		(-0.94)	(-1.46)	(-1.48)	(-1.01)	(0.36)	(0.18)	(0.14)	(0.76)	(0.33)
	FXB	1.04	1.03	1.00	1.01	0.99	8.68	26.94**	0.80	8.91
		(1.16)	(0.44)	(0.01)	(0.04)	(0.10)	(0.12)	(0.00)	(0.98)	(0.54)
	FXS	0.96	0.87	0.84	0.82	0.99	5.98	19.49*	4.04	8.69
		(-1.18)	(-1.70)	(-1.38)	(-1.06)	(0.24)	(0.31)	(0.03)	(0.54)	(0.56)
	CNY	0.70	0.40*	0.24*	0.18*	0.97**	157.97**	181.37**	25.42**	30.33**
		(-1.89)	(-2.22)	(-2.33)	(-2.05)	(0.00)	(0.00)	(0.00)	(0.00)	(0.00)
	INR	0.85**	0.64**	0.57**	0.52*	0.99*	45.62**	52.13**	24.81**	35.92**
		(-2.67)	(-3.48)	(-3.12)	(-2.50)	(0.03)	(0.00)	(0.00)	(0.00)	(0.00)

Table 1.24: The test statistics of 9 currency ETFs for the period Jan 1, 2008 - Sep 30, 2013. The variance ratios $1+M(q)$, $q = 2, 5, 10, 20$, are reported in VR labeled columns and in the main rows in line with the ETF tickers, whereas the corresponding z-scores are shown in parentheses immediately below each main row. The DF labeled column shows the first lag regression coefficients of log prices for the Dickey-Fuller test in the main rows with the corresponding p-values in parentheses immediately below the main rows. The LB labeled columns represent the Q-statistics for the Ljung-Box test on the innovations w_k in the main rows with p-values in parentheses immediately below each main row. Similarly, the GLB labeled columns show the Q-statistics for the Ljung-Box test on the standardized residual e_k from fitting returns to a GARCH(1,1) model. Significance levels at 99% and 95% are signaled by two and one stars next to the main row statistics respectively. Under the random walk null hypothesis, the value of the variance ratio is 1, the Dickey-Fuller regression first lag coefficient is 1 and w_t's are independent. Under the random walk with GARCH residuals price hypothesis, the Dickey-Fuller regression first lag coefficient is 1, LB Q-statistics are significant and GLB Q-statistics are insignificant.

ports of commodities and goods boosting the countries growth. Potentially, as a by-product of this control, the managed currencies show statistically significant negative autocorrelations, exhibiting mean-reverting behaviors for the sample period we considered.

Another factor which can participate in explaining our observation is that the structures of these products are somewhat different. For instance, FXE is a trust of the developed currencies, which holds euro and sometimes issues and redeems baskets in exchange of deposits of euro. Its investment objective is for its shares to reflect the price in USD of the euro. If the euro appreciates relative to the USD and a shareholder sells the FXE shares, he or she will earn income, otherwise, incur a loss. As to CNY, it does not directly invest in the Chinese Renminbi (RMB) as the Chinese government restricts spot transactions in RMB. Instead, it is structured as a senior unsecured debt which is based on the the S&P Chinese Renminbi Total Return Index. It offers investors the opportunity to receive at maturity or upon certain early requests an amount of cash which may be more or less than the stated principal amount based on the positive or negative performance of the index. The index is designed to be an investable alternative to the spot exchange rate between USD and RMB and tracks the value of rolling three-month non-deliverable currency forward contracts. Unlike typical debt securities, CNY does not pay interest and do not guarantee any return of principal at maturity.

1.4 Conclusion

In this chapter, we have investigated the price dynamics of a collection of major ETFs in the US market, whose underlying assets range over equities, bonds, commodities and currencies. We find that for the sample period January 2006-September 2013, the non-random-walk behaviors primarily reside in the ETFs which track or invest in large-cap, value-stock based equity indices, short-term Treasury debt indices and emerging market currencies. Little evidence against the random walk model is observed in the commodity ETFs, except for the natural gas exchanged trade fund, UNG. We have also tested an alternative return-generating process in Section 1.2.4. The results show that the model is capable of explaining the price dynamics of a great portion of the ETFs we have considered. The rejection of this model occurs mostly in the funds whose returns exhibit strong serial correlations.

The non-random walk behaviors in ETFs have been seen often affected by the relative valuation, the liquidity of their underlying assets and the various governmental policies adopted at different times in the economic cycle. For equity ETFs, we identify size and style to be two essential factors affecting the magnitude of the variance ratio statistics. Equity broad market index ETFs with large dividend yields, book-to-market ratios and market capitalizations tend to exhibit less-than-one variance ratios (or negative return serial correlations) despite the heteroscedasticity and non-normality in their returns with high statistical significance especially during low interest rate and high volatility environments. A collection of the sector ETFs in the US market have been evaluated to confirm this observation.

The short-term[25] Treasury debt based ETFs we have investigated display strong rejections to the random walk hypothesis due to negative autocorrelations in their returns, whereas funds based on Treasury notes or bonds with longer maturities do not. In the corporate bond ETFs, we discover discrepancies between the price dynamics of the ETFs and those of their underlying indices. We justify this phenomenon by the illiquidity of the corporate bonds making up those indices and the structural property of the exchange traded funds which allows them to trade at either discount or premium to their net asset values. We also document evidence which may suggest the existence of a lead-lag effect between the ETFs and their underlying bond markets by investigating into a selected number of high yield mutual funds. We find that the returns of HYG lead those of the high yield mutual funds.

UNG is the only commodity ETF showing significant levels of serial correlations among the ones researched in Section 1.3.3. We build a simple mathematical model to study the effects of the rolling mechanism and the term structure of the natural gas futures on the ETF since its inception. We find that the low levels of autocorrelations in the daily returns of UNG tend to coincide with the term structure of the natural gas futures being in

[25]Here, the short-term Treasuries refer to those whose maturities range from 1 month to 3 years.

contango. In addition, during the sample period, the historically observed extreme co-movements between natural gas and crude oil no longer exist. Natural gas experienced oversupply and its dollars per million BTUs have become less than one third of those of the crude oil, whereas prior to 2005 the dollars per million BTUs of natural gas and crude oil were at comparable levels.

From the currency ETFs, we observe that foreign exchange policies may play an important role. The ETFs whose underlying currencies are subject to more central planning tend to have more predictability, leading to the rejection of the random walk hypothesis. The prices of the ETFs based on free-floating currencies, however, show dynamics in consistency with the random walk model.

Our results in this chapter tend to support the dictum of Paul Samuelson, who believed that financial markets are generally micro-efficient but macro-inefficient[26], as the statistically significant predictabilities we have found are mostly in those ETFs which track broad market indices. However, the rejection of the random walk hypothesis, which is purely statistical, by no means equates with market inefficiency. As LeRoy (1973) and many others have rigorously shown, unpredictable prices may not imply a well-functioning market with rational participants, and vice versa. More generally speaking, although market participants often trade based on information, factors such as varying risk aversion, liquidation constraints etc. can incur transactions, as well. Investors who are forced to liquidate portions or the entirety of their portfolios can certainly offer gains to those who are well-capitalized. Yet, each side of those trades make the necessary decisions to maximize their own utilities, neither irrationally nor inefficiently. Nonetheless, the non-random-walk behaviors may as well offer certain minority groups of investors the opportunity to achieve additional profits. Provided with sufficient liquidity, ETFs whose returns show statistical significant less-than-one variance ratios may be suitable for certain contrarian investment strategies, whereas those with considerably greater-than-one variance ratios for trend-following ones. Above all, we brace the

[26]What this means is that the minority of investors who spot aberrations derived from certain fundamental measurements on individual securities can profit from those occurrences and, therefore, wipe out any persistent inefficiencies. However, when individual securities are bundled into indices, these measurements in aggregate may not work any more. As a result macro-inefficiencies can continue to exist.

concept that markets are efficient when subject to few capital controls, regulatory restrictions or uneven financing costs. Violations of such ideal circumstances are likely to offer opportunities for market participants to achieve excess returns.

1.5 Appendix for Chapter 1

1.5.1 *Proof of Proposition 1.1*

Recall that the price of the ETF denoted by P_t can be represented in the following form

$$P_t = P_{t-1}\frac{(1-w_{t-1})F_t + w_{t-1}B_t}{(1-w_{t-1})F_{t-1} + w_{t-1}B_{t-1}}.$$

By multiplying the numerator and denominator by F_t and F_{t-1}, we may obtain the next equation by rearranging the terms,

$$\begin{aligned} P_t = P_{t-1}&\frac{(1-w_{t-1})F_t + w_{t-1}B_t}{(1-w_{t-1})F_t} \times \frac{F_t}{F_{t-1}} \\ &\times \frac{(1-w_{t-1})F_{t-1}}{(1-w_{t-1})F_{t-1} + w_{t-1}B_{t-1}}. \end{aligned} \tag{1.49}$$

If we denote the slope of the term structure of the futures plus 1 as

$$s_t = \frac{B_t}{F_t}, \tag{1.50}$$

then it follows that

$$s_t = e^{c_t/12} \tag{1.51}$$

and Eq. (1.49) can be rewritten as

$$\frac{P_t}{P_{t-1}} = \frac{(1-w_{t-1}) + w_{t-1}s_t}{(1-w_{t-1}) + w_{t-1}s_{t-1}} \cdot \frac{F_t}{F_{t-1}}. \tag{1.52}$$

Taking the logarithm on both sides of this equation, we get the return of the ETF, p_t, as follows

$$p_t = \log\left(1 + w_{t-1}\frac{s_t - s_{t-1}}{(1-w_{t-1}) + w_{t-1}s_{t-1}}\right) + \log\frac{F_t}{F_{t-1}}$$

$$
\begin{aligned}
&= \frac{w_{t-1}\Delta s_t}{1+(s_{t-1}-1)w_{t-1}} + \log\frac{F_t}{F_{t-1}} + O\left(w_{t-1}^2(\Delta s_t)^2\right) \\
&= \frac{w_{t-1}\Delta s_t}{1+(s_{t-1}-1)w_{t-1}} + \Delta c_t\tau + c_{t-1}/252 + r_t + O\left(w_{t-1}^2(\Delta s_t)^2\right) \\
&= \frac{w_{t-1}\Delta s_t}{1+(s_{t-1}-1)w_{t-1}} + 12\Delta(\log s_t)\tau + \frac{c_{t-1}}{252} + r_t + O\left(w_{t-1}^2(\Delta s_t)^2\right) \\
&= \left(\frac{w_{t-1}}{1+(s_{t-1}-1)w_{t-1}} + \frac{12\tau}{s_{t-1}}\right)\Delta s_t + \frac{c_{t-1}}{252} + r_t + O\left(\left[w_{t-1}^2 \vee 12\tau\right](\Delta s_t)^2\right)
\end{aligned}
$$

where

$$
\frac{F_t}{F_{t-1}} = \frac{S_t}{S_{t-1}} e^{\Delta c_t\tau + c_{t-1}/252} \tag{1.53}
$$

and r_t is the log return of the spot price S_t.

Chapter 2

Models of Return Autocorrelation

2.1 Introduction

In this chapter, we will develop models which help explain how various factors such as size, style, return volatility, interest rate, flow of information, trade size and trading cost influence the serial correlation of asset returns. Some aspects of the models detailed here provide justifications for the empirical results obtained on ETFs and portfolios of securities in Chapter 1, others will introduce new insights into the behaviors of return autocorrelation. The results in this chapter can also be viewed as extensions to existing research on return autocorrelation such as Lo and MacKinlay (1988), Roll (1984), Campbell, Grossman and Wang (1993) and Almgren and Chriss (2001), etc. We will start by providing some interesting background and development on this topic.

2.1.1 *Bid-ask bounce*

Sources of return autocorrelation can vary. Some researchers show that microstructure effects can induce spurious return autocorrelation. For instance, Roll (1984), in the course of obtaining a measure of effective bid-ask spreads, found that if the asset price is assumed to follow a random walk, p_t, but trade at either ask price, $p_t + c$, or bid price, $p_t - c$, with a constant transaction cost c (so that the bid-ask spread is 2c), then returns[1] com-

[1]Returns considered by Roll are simply price differences.

puted using the trade prices exhibit negative autocorrelation, even when the true price returns based on p_t has in theory zero autocorrelation.

More specifically, suppose that the true price p_t follows

$$p_t = p_{t-1} + u_t, \tag{2.1}$$

where u_t is an independent random noise which has a constant standard deviation σ_u. The realized prices denoted by m_t (or the bid and ask prices) will then follow

$$m_t = p_t + cq_t, \tag{2.2}$$

where q_t can take values either 1 or -1 with equal probability and q_t is independent of u_t.

Given this dynamic, we can easily write the change in price, Δm_t, as

$$\begin{aligned} \Delta m_t &= m_t - m_{t-1} \\ &= -cq_{t-1} + cq_t + u_t, \end{aligned}$$

and compute its variance as

$$Var(\Delta m_t) = 2c^2 + \sigma_u^2 \tag{2.3}$$

and its covariance between consecutive price changes as

$$Cov(\Delta m_t, \Delta m_{t-1}) = -c^2. \tag{2.4}$$

The first lag serial correlation will therefore be

$$Corr(\Delta m_t, \Delta m_{t-1}) = \frac{-c^2}{2c^2 + \sigma_u^2}. \tag{2.5}$$

As a result, under this naive assumption, we see that due to bid-ask bounce, the returns tend to be negatively autocorrelated when they are realized in the hypothesized market.

The immediate question on Roll's model is - does this method fits the empirical data? Harris (1990) finds that in practice, lots of empirical tests show price changes to be positively serial correlated rather than negatively, leading to the failure of computing expected bid-ask spreads when the square root is taken. He demonstrated that Roll's estimates are"severely biased by Jensen's inequality". In particular, daily estimates are smaller than weekly estimates.

In addition, Roll's model can be overly simplistic in many aspects of its assumptions. For instance, if we take order sizes into consideration, then it is quite reasonable to assume bid prices to be more frequently followed by bid prices rather than completely random when a large order is being executed. Bid-ask spread is not constant in time, either. Microstructure research shows that bid-ask spreads are functions of trading volume and asset return volatility. That is, securities which experience large trading volumes and low return volatility tend to have small bid-ask spreads. Further, this model does not account for the different roles played by the market makers and market takers, as a result of which, those players may be subject to information asymmetry.

Although Roll model does not appear to be empirically successful, it is still often regarded as the first market microstructure model due to its simplicity. Many extensions of this model were created by other authors. For instance, Choi et al. (1988) incorporate the possibility of autocorrelation in transaction types. Glosten & Milgrom (1985) made further progress in the study of the bid, ask and transaction prices by introducing information asymmetry in the market participants rather than them being homogeneous. According to their model, whether the bid-ask spread will cause negative autocorrelation depends on its source. That is, if the bid-ask spread is solely caused by adverse selection[2], then it will not induce serial correlation in price differences. However, if the bid-ask spread is caused by factors such as specialist transaction costs, risk aversion or monopoly power, then it will give rise to negative autocorrelation in price differences.

2.1.2 *Non-synchronous trading*

When stocks do not trade, old quoted prices remain in effect and are recorded as if they are current prices. Those old prices are called stale prices. Stale prices will not reflect the recent information and directly linked to the liquidity of the asset. If a portfolio is made up of infrequently

[2]The paper assumes a specialist through whom customers buy and sell certain financial assets. The adverse selection comes from the idea that a customer agreeing to trade at the specialist's bid or ask price may be trading because of information that is unknown to the specialist. As a result, the specialist will have to recoup his losses in transactions with the informed traders by gains with liquidity traders or non-informational traders.

traded stocks, they will not trade simultaneously, which leads to the phenomenon of non-synchronous trading. The second line of autocorrelation research apart from the bid-ask bounce focuses on how non-trading affects the serial correlation of individual stocks and portfolios of securities.

Among the authors in this angle of research, Lo and MacKinlay (1988) found empirically that individual stock returns show weak levels of negative autocorrelation but portfolio returns of individual stocks display strong positive autocorrelation using stocks from the CRSP database. In addition, they developed a non-synchronous trading model which can accommodate these empirical behaviors.

Essentially, in their work, they first define a process for the true prices of assets and a separate process for the realized prices, where the realized prices are a function of the true prices and a probability which measures how often stocks do not trade. Based on this price dynamics, a closed form solution of the first-lag serial correlation of portfolio returns is obtained as a function of the non-trading probability for both individual asset and portfolios of assets. By using this formula, the authors were able to deduce the level of theoretical autocorrelation from the empirical non-trading probability. When they compare the level of the model implied autocorrelation with the realized serial correlation computed on the portfolio returns, they observe that the sample autocorrelations for size-sorted portfolios are much larger than what the simple model implies. For instance, for a large cap portfolio of US stocks they considered, the sample first-lag autocorrelation of daily returns from 1973-1987 is 0.165, yet the model implied autocorrelation from non-trading is only 0.008. As a result, it is believed that non-synchronous trading cannot be the only factor in explaining the return autocorrelation. Other authors such as Mech (1993) also have similar findings that non-trading is not sufficient to explain portfolio return autocorrelation alone.

2.1.3 *Time-varying risk aversion*

Another group of researchers such as Campbell, Grossman and Wang (1993) [Campbell *et al.* (1993)] investigated the effects of trading volume on individual stock returns through the time-varying risk aversions of different market participants. Their economic equilibrium model, in which risk-averse market makers accommodate buying or selling pressure from liquidity/ non-informational traders, suggests that the change of risk aversion of the marginal investors of the economy justifies the expected level of future returns and induces trading volume. Their model also implies that a drop in asset prices on days of high trading volumes tend to be more likely followed by an increase in the expected return in subsequent days than a price decline on low volume days.

Avramov et al. (2006) continued along this direction by showing a strong relationship between short-run reversals and stock illiquidity, even when trading volume is controlled. Empirically, they studied a sample of NYSE-AMEX stocks for 1962-2002 and found that there were reversals in short-run stock returns, mostly concentrated in loser stocks. In particular, at the weekly frequency, high turnover, low liquidity stocks exhibit stronger negative serial correlation than low turnover, high liquidity stocks. From a theoretical perspective, they hypothesized that the slope of the demand curve should be steeper for illiquid stocks, which add to their price reversals upon non-informational liquidity trading. However, they argue in favor of market efficiency because the profits from the contrarian strategies they tested were not large enough to contain the necessary transaction costs involved.

2.1.4 *Partial price adjustment*

As more and more financial data in the stock market became available, an increasing number of research has been able to separate effects like bid-ask bounce and non-trading, and concluded that the most likely reason for serial correlation is partial price adjustment. For instance, Anderson et al. (2012) decomposed stock return autocorrelation into spurious components (bid-ask bounce and non-trading) and genuine components (partial price adjustment and time-varying risk premia). Some of their techniques include computing returns over disjoint sub-periods separated by a trade using intraday data to reduce non-synchronous trading and bid-ask bounce; and dividing the data periods into disjoint sub-periods to obtain independence for statistical power.

Their empirical testing using 16 years of NYSE intraday transaction data supports partial price adjustment as a major source of autocorrelation. Further, they showed partial price adjustment may have both positive and negative effects on return autocorrelation. Positive autocorrelation can be caused by the strategic decision of informed traders who exercise their informational advantage slowly; negative autocorrelation may result from un-informed traders attempting to exploit the information of the informed traders using momentum strategies.[3]

The rest of this chapter will discuss two additional models in detail. The first one extends the non-synchronous trading model by Lo and MacKinlay. In their research, they assumed the returns of individual securities to satisfy a one-factor model, with the factor being an independently and identically distributed random variable, so that the non-trading effect is singled out. Empirically, we have found that several well-known market factors actually exhibit significant levels of autocorrelation for the period 1926-2013 and their levels shift with rather wide ranges. Therefore, we attempt to incorporate this factor autocorrelation element into the existing framework and see how it translates into the return autocorrelation of a well-diversified portfolio together with the non-trading effect. Section 2.2 illustrates this model and shows that the resultant portfolio first-lag return autocorrelation is equal to the original non-trading probability plus a multiple of the factor autocorrelation.

In Section 2.3, we describe an economic model based on the interaction between risk-averse market makers and market takers. Their trading activities of the economy's risky asset result in a relation between the asset's return autocorrelation and its average trade size. We find that return autocorrelation is inversely related to the average trade size of an individual security and empirical experiments performed using NASDAQ daily data confirm this model implication. The model also provides several additional implications in regard with the effects of book-to-price ratio, interest rates, insider trading and transaction costs on return autocorrelation.

[3]These can also be interpreted as the effects of feedback trading, a simple model for which was developed by Sentana and Wadhwani in 1992.

2.2 Non-synchronous trading model with autocorrelated factors

We consider a market consisting of individual stocks which may experience infrequent trading. When a stock does not trade, its price remains constant and will not reflect the most recent information content in the market. To account for this delayed information capture, we define two related forms of returns, intrinsic[4] and realized, for each stock. The intrinsic stock returns are denoted by r_{it} for stock i at time t, which will follow a one-factor model of the form,

$$r_{it} = \mu_i + \beta_i f_t + e_{it},$$

where e_{it} are independently and identically distributed (iid) random variables with mean 0 and variance σ_i^2. f_t is the market factor and is assumed to be autocorrelated via the following expression

$$f_t = \alpha f_{t-1} + w_t,$$

where $|\alpha| < 1$, w_t is iid with mean 0 and variance σ_f^2 and uncorrelated with e_{it}. It is latent and not observable by the market participants. The realized returns are represented by r_{it}^o and they can be different from the assumed dynamic of r_{it}. For instance, on a day of no trading activities of the stock i, the price of the stock will remain the same; therefore, we expect r_{it}^o to be 0. However, the intrinsic returns are assumed to have already accounted the information accumulated in the day, so that r_{it} will vary based on f_t and e_{it}.

To better represent the relationship between the two forms of returns. We suppose at each time t, non-trading happens with a probability π_i. Further we assume that π_i is independent of the stock's intrinsic return r_{it} as well as all other random variables we consider in this model. Then, an indicator function of non-trading at time t denoted by δ_{it} can be defined such that

$$\delta_{it} = 1 \text{ with probability } \pi_i \tag{2.6}$$

and

$$\delta_{it} = 0 \text{ with probability } 1 - \pi_i. \tag{2.7}$$

[4]The intrinsic returns here are sometimes also referred to as virtual returns.

Therefore, the event that stock i is traded at time t but has not been traded for the past k periods can be represented by $(1-\pi_i)\pi_i^k$, for any $k>0$. We may in fact define an indicator function, $X_{it}(k)$, of such event using δ_{it}, as well. That is, $X_{it}(k)$ can be written as a product of δ_{it}'s in the form of

$$X_{it}(k)=(1-\delta_{it})\delta_{it-1}\delta_{it-2}...\delta_{it-k}.$$

Now, it is easy to see that $X_{it}(k)=1$ with probability $(1-\pi_i)\pi_i^k$ and 0 otherwise, where $k>0$. When $k=0$, naturally, $X_{it}(0)=1-\delta_{it}$.

Hence, the realized return r_{it}^o can be represented by the following equation,

$$r_{it}^o=\sum_{k=0}^{\infty}X_{it}(k)r_{it-k}.$$

More explicitly, this is equivalent to

$$r_{it}^o=0 \text{ with probability } \pi_i$$

and

$$r_{it}^o=r_{it}+...+r_{it-k} \text{ with probability } (1-\pi_i)^2\pi_i^k$$

for $k>0$.

What this says essentially is that if the stock was traded two days ago but did not trade yesterday, then when it trades today, not only does it need to reflect today's intrinsic return r_{it}, it also has to reflect yesterday's intrinsic return r_{it-1}, therefore, what observed today will be $r_{it}^o=r_{it}+r_{it-1}$, reflecting the information for both days. The following table illustrates the relationship between the observed returns and the intrinsic ones.

r_{it}^o	t	$t-1$	$t-2$
0	not trade		
r_{it}	trade	trade	
$r_{it}+r_{it-1}$	trade	not trade	trade

Table 2.1: The relationship between the observed returns and the virtual or latent ones.

Based on this formulation, it is almost apparent that realized returns may exhibit spurious autocorrelation, since future observed returns may have a chance to depend on past returns. To show how the model parameters affect the serial dependence of realized returns, we compute their auto-covariance and cross-covariance. The results are summarized in the following proposition.

Prop 2.1. Suppose the intrinsic return of an individual stock follows

$$r_{it} = \mu_i + \beta_i f_t + e_{it} \tag{2.8}$$

with

$$f_t = \alpha f_{t-1} + w_t, \tag{2.9}$$

where w_t is iid with mean 0 and variance σ_f^2 and uncorrelated with e_{it}, which has mean 0 and variance σ_i^2. If we denote

$$X_{it}(k) = (1-\delta_{it})\delta_{it-1}\delta_{it-2}...\delta_{it-k}, \tag{2.10}$$

then the realized return r_{it}^o can be defined as

$$r_{it}^o = \sum_{k=0}^{\infty} X_{it}(k) r_{it-k} \tag{2.11}$$

and its mean, variance, auto-covariance and cross-covariance can be computed as

$$E(r_{it}^o) = \mu_i \tag{2.12}$$

$$Var(r_{it}^o) = \frac{\beta_i^2\sigma_f^2}{1-\alpha^2} + \sigma_i^2 + \frac{2\pi_i}{1-\pi_i}\mu_i^2 + \frac{2\beta_i^2\sigma_f^2\alpha\pi_i}{(1-\alpha^2)(1-\pi_i\alpha)} \tag{2.13}$$

$$Cov(r_{it}^o, r_{it+1}^o) = -\pi_i\mu_i^2 + \frac{(1-\pi_i)^2\beta_i^2\sigma_f^2\alpha}{(1-\alpha^2)(1-\alpha\pi_i)} \tag{2.14}$$

$$Cov(r_{it}^o, r_{jt+1}^o) = \frac{\beta_i\beta_j(1-\pi_i)(1-\pi_j)}{1-\pi_i\pi_j} \times \frac{\sigma_f^2}{1-\alpha^2} \times \left\{\pi_j + \frac{\alpha\pi_j^2}{1-\pi_j\alpha} + \frac{\alpha}{1-\pi_i\alpha}\right\}. \tag{2.15}$$

The proof of this proposition is in the appendix of this chapter.

From this proposition, we see that α participates in all second moments of the realized returns. If $\alpha > 0$, then the auto-covariance of r_{it}^o will be larger than when the factor f_t is uncorrelated, since the term

$\frac{(1-\pi_i)^2\beta_i^2\sigma_f^2\alpha}{(1-\alpha^2)(1-\alpha\pi_i)} > 0$. So, the temporal dependence of the factor is passed onto that of the individual security i. The cross-covariance are affected in a similar fashion as well. If $\alpha < 0$, which may suggest a distressed market environment if f_t represents the market factor, then we see the auto-covariance becomes more negative than $-\pi_i\mu_i^2$; but the cross-covariance might remain positive if the non-trading probability exceeds the effect of α.

With the statistics computed for individual securities, we continue to compute those for portfolio of securities. The next proposition will show that the autocorrelation of the factor will show up as an important component in the portfolio autocorrelation.

Prop 2.2. If we consider an equally weighted portfolio of N securities with identical non-trading probability, π, and its observed return is of the form

$$r_t^o = \frac{1}{N}\sum_{i=1}^{N} r_{it}^o, \tag{2.16}$$

then its autocorrelation denoted by ρ_α will approach

$$\rho_\alpha = \frac{\pi+\alpha}{1+\pi\alpha}, \tag{2.17}$$

as $N \to \infty$.

The proof of this proposition is in the appendix of this chapter.

One interesting property of Eq. (2.17) is that π and α are symmetric in the expression of ρ_α. This might imply that non-trading can be represented via a particular factor in the return dynamic if we adopt certain multi-factor frameworks for the modeling.

Empirically, the autocorrelations of return factors are indeed changing. For instance, under Fama-French three-factor model

$$r_{it} = R_{ft} + \beta_M(R_{Mt} - R_{ft}) + \beta_s SMB_t + \beta_v HML_t + e_{it}, \tag{2.18}$$

where R_{ft} is the risk-free rate at time t, β_M, β_s and β_v are coefficients, $R_{Mt} - R_{ft}$ is market premium, SMB_t is small minus big cap factor, HML_t is high minus low book-to-price ratio factor and e_{it} is idiosyncratic innovation. We find all three factors exhibit strongly significant serial correlation.

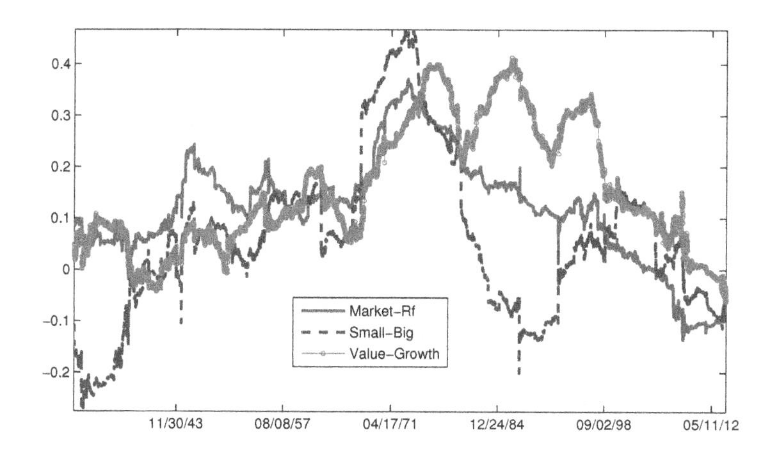

Fig. 2.1: The first-lag autocorrelations of daily returns rolling over the entire history with a window of 5 years for the portfolio of market premium and Fama-French size and style factors respectively for the period 1926-2013.

Figure 2.1 plots the first lag autocorrelations of daily returns rolling over the entire history with a window of 5 years for the portfolio of market premium and Fama-French size and style factors respectively for the period 1926-2013. We see that the levels of autocorrelation for all three factors moved rather wildly ranging from below -0.2 to over 0.4. In particular, in the most recent crisis, all factors' autocorrelations fall below zero. Together with the findings in Chapter 1 that all major broad market and sector index returns show negative autocorrelations for the same period, our results here suggest that the inclusion of factor autocorrelation into the model assumptions can better account for the dynamics of autocorrelations for portfolio returns. The predictability of indices can therefore be analyzed via a factor approach. For instance, in the 1980s, the positive autocorrelations in the broad market indices are likely to be justified by the value factor and market premium but not the size factor.

In addition, Fama-French factors remain important in explaining the magnitudes of the return autocorrelations of ETFs. Figure 2.2 plots the effects of Fama-French size and style factors on the return autocorrelations of sector ETFs for the period 2006-2013. The vertical axis of the plot represents the daily return autocorrelation of the sector ETFs for the period 2006-2013. The x-axis represents the average market capitalization of the holdings of the sector ETFs. The y-axis represents the style score defined in Chapter 1. The surface is obtained by linear interpolation of the circles representing the ETFs. We see that as the size and style score increase, the return autocorrelations of the sector ETFs decrease.

2.3 Time varying risk aversion model

In this section, we consider an economy which is made up of two types of traders: market makers and market takers. Market takers can either trade based on exogenous factors or on private information, whereas market makers serve the economy as liquidity providers, offering bid and offer quotes on market securities and profiting from the spread as well as potential trading impacts.

A general mechanism illustrating the effect of trading on return autocorrelation can be described as follows. Suppose a fall in a stock's price is observed in an economy. Three events are likely to be the cause of this decline:

(1) the arrival of new public information makes all investors in this stock lower its valuation;

(2) exogenous factors impose selling pressures of this security on liquidity or non-informational traders;

(3) insiders with negative private information about the stock secretly take actions in liquidating the security.

In the first case, we expect the volume of the stock to be small as the information is public to all investors; it should be incorporated into the stock price rather efficiently (assuming that this public information allows traders to correctly price the stock) and the stock returns should experience more volatility than serial correlation. In the second case, when market makers, who maximize their own utilities, accommodate the selling,

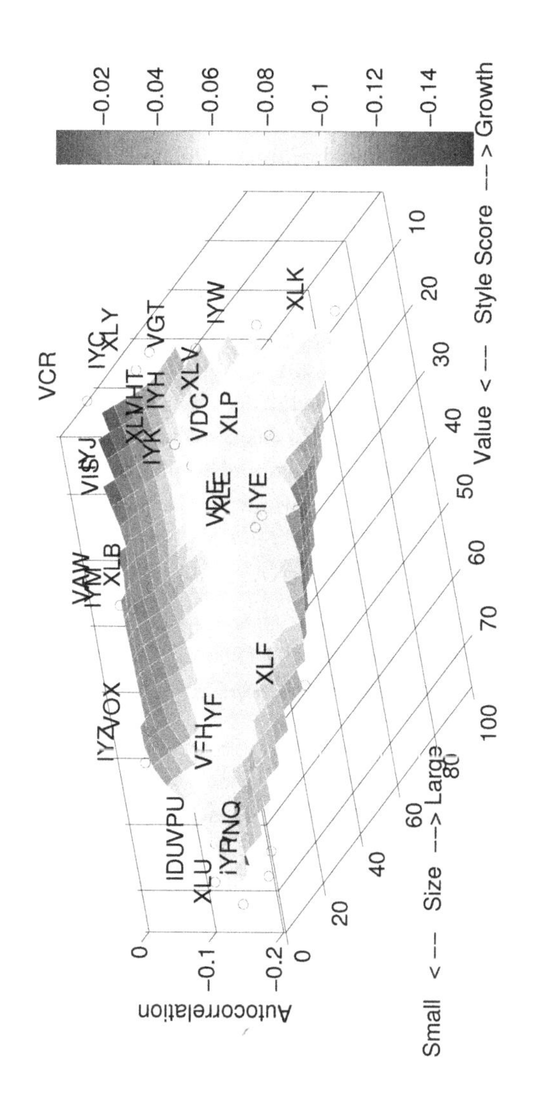

Fig. 2.2: The effects of Fama-French size and style factors on the return autocorrelations of sector ETFs for the period 2006-2013.

realizing the price decline, they also demand a reward in the form of a higher expected stock return. Consequently, the stock price is more likely to rise in the subsequent days. Price drops under selling pressure of liquidity traders are typically accompanied with unusual volume due to the digestion of exogenous information revealed in the process. Thus, a stock price fall on a high volume day is more likely than a stock price fall on a low volume day to be followed by an increase in the expected stock return [Campbell *et al.* (1993)]. In the third case, however, the price is likely to fall again in the following days, but exhibiting positive autocorrelation instead, because the insiders will continue to gradually exploit their private information, though slowly, until either the prices of the security drop to the level as their signal suggests or other traders detect their activities and follow suit.

The model we discuss in this section attempts to explain this mechanism. The rest of this section is laid out as follows. We set up the economy and price dynamics in Section 2.3.1 and describe the traders' utility maximization problem in Section 2.3.2. In Section 2.3.3, we solve the equilibrium problem, and then illustrate the model implications on how different factors affect stock return autocorrelations in Section 2.3.4. Section 2.3.5 shows empirical results on the inverse relation between the average trade size and the return autocorrelation in NASDAQ traded stocks.

2.3.1 *The economy and price dynamics*

We consider an economy of two types of assets: a risk-free asset and a risky asset or stock. The risk-free asset offers a constant return denoted by r. For simplicity, we assume that the asset financing rate is also at this level. The stock has a dividend adjusted price of P_t at time t. The total number of shares of the stock are assumed to be fixed and normalized to 1. Instead of using a dividend discount model for the stock price valuation, we adopt an accounting based approach. We set the stock price as a constant multiple, β, of the fundamentals of the company, say, its book value. That is

$$P_t = \beta B_t, \tag{2.19}$$

where B_t is the book value at time t. Naturally, β ($\beta > 0$) can be interpreted as the company's price-to-book ratio. The book value of a company is essentially the stockholder's equity level in the firm's balance sheet. In a rather simple scenario, it consists of common stock and retained earnings. If the company has no share operations such as common stock repurchase or share new issuance offerings, the common stock level remains constant. Based on the yearly or quarterly revenue generating operations of the company, the retained earnings account is updated by the net income for the period subtracting dividends, if any.

For the moment, we model the dynamics of book values by the following equations,

$$B_t = \bar{B} + \tilde{B}_t, \tag{2.20}$$

with

$$\tilde{B}_t = \alpha \tilde{B}_{t-1} + u_t, \tag{2.21}$$

where u_t is identically and independently distributed normal random variable with mean 0 and variance σ^2, and $|\alpha| < 1$. $\bar{B}$ is the mean level of B_t and $\tilde{B}_t$ follows a stationary AR(1) process. For equity index ETFs, this can be referred to as the average of the book values of the constituents of the fund holdings. Note that this assumption can be too simplified to meet reality for individual stocks as fundamental values of a firm are often announced quarterly. In addition, upon announcements of these accounting numbers, sizable impacts may take place in trading, reflecting reprice of the market values of the stocks. Therefore, our setting may suit pooled investments better, as idiosyncratic risks are more or less diversified. Also, since we seek to learn the behavior of short term autocorrelation of asset returns, we assume the fundamentals to mean revert. A linear time trend can be added to the dynamics of B_t, but it does not affect return autocorrelation, thus, for simplicity, this additional assumption is omitted.

According to the above assumptions, the price dynamics will follow

$$P_t = (1 - \alpha)\beta \bar{B} + \alpha P_{t-1} + \beta u_t, \tag{2.22}$$

with innovations being identically and independently distributed normal random variables with mean 0 and variance $\beta^2\sigma^2$. In addition, we may define the excess return of the stock return at time t by Q_t,

$$Q_{t+1} = P_{t+1} - (1 + r)P_t, \tag{2.23}$$

where r represents the financing costs of the security.

2.3.2 *The traders and their utility problems*

The traders of this economy are assumed to be mainly of two types: market makers and market takers. They are both risk-averse; they determine how many shares of stocks to hold by maximizing their own utility functions. The utility functions for the traders are defined to be the mean of their wealth subtracted by a multiple of risk as in Eq. (2.24). λ is used to denote the level of risk aversion. The market takers can be influenced by either exogenous factors or private information, for which we assign them time-varying risk-aversion b_t; market makers on the other hand are assumed to have a constant risk-aversion level denoted by a ($a > 0$). Let w be the fraction of market markers, then $1 - w$ is for market takers.

In each period, the traders solve the following problem:

$$\max_{X_t} E_t W_{t+1} - \frac{\lambda}{2} Var_t W_{t+1} + s(\lambda)\frac{c_\lambda}{2}(X_t - X_{t-1})^2 \tag{2.24}$$

subject to

$$W_{t+1} = W_t(1+r) + X_t[P_{t+1} - (1+r)P_t], \tag{2.25}$$

where X_t is the number of shares of the stock the trader chooses to hold to maximize his utility, W_t represents the wealth of the trader at time t and c_λ is trading cost per share. Equation (2.25) defines the dynamics of the wealth of the trader from period t to period $t+1$. That is, the $t+1$ period wealth equals the sum of the original wealth W_t, the return on this wealth invested in the risk-free asset rW_t and the gain from investing X_t shares of the stock. We have assumed that the traders can finance the purchase of the stocks at rate r. The function $s(\lambda)$ is a sign function such that

$$s(\lambda) = \begin{cases} 1, \lambda = a, \\ -1, \lambda = b_t, \end{cases}$$

where we have assumed that the cost is paid by market takers to market makers, thus a cost to the market takers but a profit to the market maker. Such quadratic cost assumption is consistent with the linear market impact model as in Almgren and Chriss (2001) and our market clearing condition discussed in the next section. It is adopted also for tractability. Through this cost term in the objective function, we also implicitly assume that the market impact is caused by the trading actions of the market takers.[5]

[5]Indeed, we can also formulate this problem using a temporarily impacted price $\tilde{P}_t$ by $\tilde{P}_t = P_t + \frac{c_b}{2}(X_t - X_{t-1})$ as in [Almgren and Chriss (2001)], together with a wealth dynamic whose gain in each period is based on the actually transacted price. The result obtained is consistent with our current problem setting.

Using the dynamics of P_t, we can solve the utility optimization problem, provided $\lambda Var_t(Q_{t+1}) > s(\lambda)c_\lambda$.

Prop 2.3. The utility maximization problem represented by Eq.(2.24) has the solution

$$X_t^\lambda = \frac{(1-R)\beta\bar{B} + (\alpha - R)\beta\tilde{B}_t - s(\lambda)c_\lambda X_{t-1}^\lambda}{\lambda\beta^2\sigma^2 - s(\lambda)c_\lambda}. \tag{2.26}$$

More explicitly, this can be expanded somewhat differently for the two types of market participants. For the liquidity traders,

$$X_t^b = \frac{(1-R)\beta\bar{B} + (\alpha - R)\beta\tilde{B}_t + c_b X_{t-1}^b}{b_t\beta^2\sigma^2 + c_b} \tag{2.27}$$

and for the market makers

$$X_t^a = \frac{(1-R)\beta\bar{B} + (\alpha - R)\beta\tilde{B}_t - c_a X_{t-1}^a}{a\beta^2\sigma^2 - c_a}. \tag{2.28}$$

In particular, if we assume $X_0 = 0$, then we may solve the above recurrence relation and obtain

$$X_t^a = \frac{-r\bar{B}}{a\beta\sigma^2}\left[1 - \left(\frac{1}{1 - a\beta^2\sigma^2/c}\right)^t\right] - \frac{\beta(\alpha - R)}{c - a\beta^2\sigma^2}\sum_{j=1}^{t}\left(\frac{1}{1 - a\beta^2\sigma^2/c}\right)^{t-j}\tilde{B}_j. \tag{2.29}$$

The proof of this proposition is in the appendix of this chapter.

2.3.3 *The equilibrium*

In this section, we define the risk aversion of the marginal investors and analyze the average trading size, asset price dynamics and excess returns as its functions.

2.3.3.1 *Marginal risk aversion*

We assume that upon market clearing, the total number of shares of the stock traded for each period is constant and normalized to 1. Thus, the corresponding market clearing condition[6] is

$$wX_t^a + (1-w)X_t^b = 1. \tag{2.30}$$

In the case where $c_\lambda = 0$, we substitute Eq. (2.26) and get in equilibrium that

$$\left(\frac{w}{a} + \frac{1-w}{b_t}\right) \frac{(1-R)\bar{B} + (\alpha - R)\tilde{B}_t}{\beta\sigma^2} = 1. \tag{2.31}$$

If we define Z_t by

$$\frac{1}{Z_t} = \frac{w}{a} + \frac{1-w}{b_t}, \tag{2.32}$$

then we may interpret it as the risk aversion of the marginal investors in the market as in [Campbell *et al.* (1993)], since Z_t is a weighted average of the risk aversion of both market makers and market takers. If the proportion w of market makers is tiny, then the change in the risk aversion of the marginal investors is pretty much the same as the change in the risk aversion of the market takers. Nonetheless, the risk aversion of the market makers is assumed constant, thus changes in the marginal risk aversion is driven by the changes in the market takers' risk appetite.

By substitution of this definition into the previous equation, we get

$$Z_t = \frac{-r\bar{B} + (\alpha - R)\tilde{B}_t}{\beta\sigma^2}. \tag{2.33}$$

Therefore, there is a one-to-one correspondence between the marginal risk aversion Z_t and the book values $\tilde{B}_t$. Under the assumption that β and w are both constant, this equation forces Z_t (or essentially b_t) to be stochastic. If we assume Z_t to be a stationary process with $Z_t = \bar{Z} + \tilde{Z}_t$ and

$$\tilde{Z}_t = \alpha_Z \tilde{Z}_{t-1} + h_t, \tag{2.34}$$

where h_t is an independently and identically distributed normal random variable with mean 0, then we must have

$$\alpha_Z = \alpha, \tag{2.35}$$

$$\bar{Z} = \frac{-r\bar{B}}{\sigma^2\beta} \tag{2.36}$$

[6]If we assume also that the total costs paid by the market takers are the same as the costs received by the market takers, then we must have $wc_a = (1-w)c_b$.

and

$$Var(h_t) = \frac{(R-\alpha)^2}{\sigma^2\beta^2}. \tag{2.37}$$

Therefore, in this simple model, we may interpret that the dynamics of $\tilde{B}_t$ is induced by the dynamics of the risk aversion of the marginal investors in the market through their interaction.

Hence, the optimal X_t^λ can be represented as

$$X_t^\lambda = \frac{Z_t}{\lambda}. \tag{2.38}$$

The equilibrium stock prices are

$$P_t = \frac{\beta(1-\alpha)}{R-\alpha}\bar{B} - \frac{\sigma^2\beta^2}{R-\alpha}Z_t \tag{2.39}$$

and the excess return can be represented by

$$Q_{t+1} = \sigma^2\beta^2 Z_t + \beta u_{t+1}. \tag{2.40}$$

Even though we cannot observe Z_t directly, it helps us considerably to understand price dynamics. For instance, we can see from the above equations that price levels are inversely associated with the risk aversion of the marginal investors. A fall in asset price can be interpreted as a rise in the market risk aversion. Furthermore, Z_t predicts a stock's next period excess returns, since the conditional expectation of R_{t+1} is

$$E_t Q_{t+1} = \sigma^2\beta^2 Z_t. \tag{2.41}$$

When Z_t is high, the market makers become much more risk-averse. Selling pressures tend to result in higher expected excess return in the following period so that the market makers when taking the opposite positions of the market takers are compensated in the subsequent periods.

Note that when nontrivial costs are considered, they are part of the expression of X_t^λ as well. In particular,

$$X_t^\lambda = \frac{Z_t + \frac{s(\lambda)c_\lambda}{\sigma^2\beta^2}X_{t-1}^\lambda}{\lambda + \frac{s(\lambda)c_\lambda}{\sigma^2\beta^2}}. \tag{2.42}$$

Therefore, both costs and risk aversion can affect the market participants' decisions on how many stocks to hold.

2.3.3.2 *Excess returns*

We have defined that excess returns $Q_t = P_t - (1+r)P_{t-1}$. If we denote $R = 1 + r$, then we have the following.

Prop 2.4. The variance of Q_t, denoted by ν, can be shown to follow

$$\nu = \frac{\sigma^2 \beta^2}{1 - \alpha^2} \left[1 + R^2 - 2R\alpha\right] \tag{2.43}$$

and its first-lag autocorrelation ρ can be expressed as

$$\rho = \frac{(R - \alpha)(R\alpha - 1)}{1 + R^2 - 2R\alpha}. \tag{2.44}$$

Moreover, ρ is increasing in α and R, with $\rho > 0$ when $\frac{1}{R} < \alpha < 1$ and $\rho < 0$ if $\alpha < \frac{1}{R}$.

The proof of this proposition is in the appendix of this chapter.

One immediate consequence of this proposition is that autocorrelations of asset returns are linked to the level of interest rates, that is, the autocorrelation increases as the interest rate increases. When interest rates R is close to 1, it is hard for the inequality $\frac{1}{R} < \alpha < 1$ to hold; in particular, if $R = 1$, the inequality becomes $1 < \alpha < 1$, which is invalid. Thus $\alpha < \frac{1}{R}$ forces ρ to be lower than 0. This is consistent with our observation in Section 1.3.1.6 of Chapter 1 on equity portfolio returns for 1926-2013. In addition, the return autocorrelations of individual stocks documented by Anderson et al. (2012) for every two-year sub-periods from 1993 to 2008 also fit this model implication. They found individual stock return autocorrelations tend to be positive pre-2000, but negative afterwards especially during the 2007-2008 sub-period.

LeBaron (1992) found that serial correlations of equity index returns are inversely related to their volatility. Our model partially justifies this empirical relation. When R is strictly greater than 1, the level of variances as in Eq. (2.43) first decreases in ρ but then increases as the autocorrelation ρ rises. This pattern is plotted in Figure 2.3. Empirically, we do find the existence of such relationship among the 100 Fama-French value-weighted portfolios constructed by size and book-to-market ratio. Take the

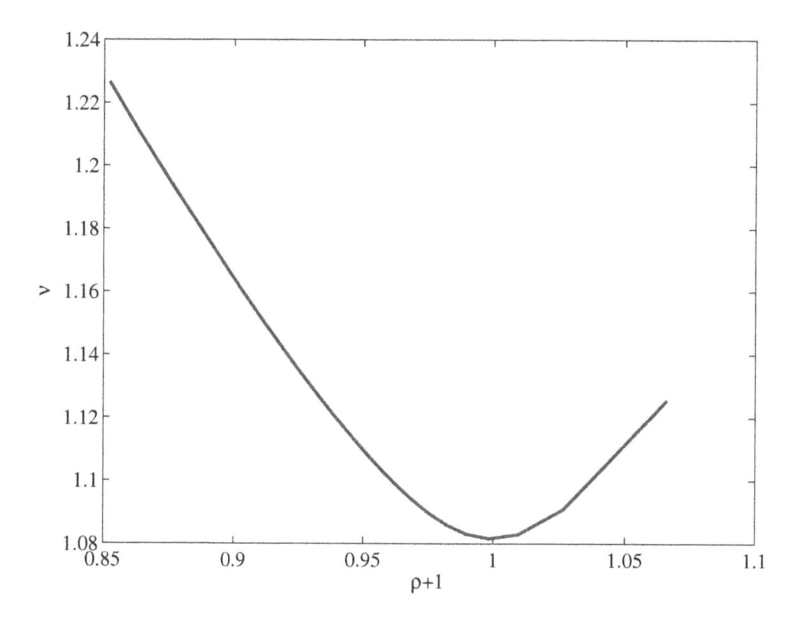

Fig. 2.3: The level of variance (ν) against the level of autocorrelation ($\rho + 1$) according to the formula in Proposition 2.4 with $\sigma = \beta = 1$.

45th portfolio for example. We use its historical daily returns of the period 1950-2013 and compute their rolling variances and autocorrelation with a window of 5 years. We then divide the variances into 20 bins and compute the average of the autocorrelations for each corresponding bins. Figure 2.4 plots the corresponding sample variances and bin-average sample autocorrelations. The blue line is fitted by a quadratic polynomial with an R^2 that is 92%. Its shape is consistent with the earlier figure generated by our model.

However, most of the 100 Fama-French portfolios show inverse relationship between variance and autocorrelation as in Figure 2.5, which represents the empirical behavior for the 56th Fama-French portfolio plotted following the same process as before. The fitted curve is in the form of a rational polynomial with numerator of degree 0 and denominator of degree 1. Its R^2 is 90%. It is not clear whether the minority quadratic behavior as shown by the 45th portfolio is due to the varying of interest rates, other economic factors or simply the discontinuity caused by the factor $1 - \alpha$ in the denom-

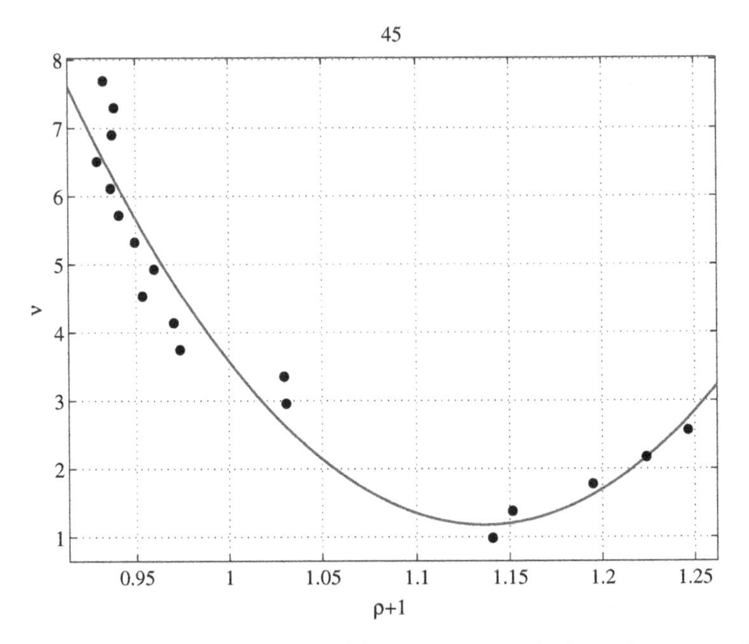

Fig. 2.4: The level of sample variance (ν) against the level of sample autocorrelation ($\rho + 1$) of the 45th portfolio daily returns of the 100 Fama-French portfolios constructed by size and book-to-market ratio. We use its historical daily returns of the period 1950-2013 and compute their rolling variances and autocorrelation with a window of 5 years. We then divide the variances into 20 bins and compute the average of the autocorrelations for each corresponding bins.

inator of the variance formula in Eq. (2.43). For the rest of the chapter, we will adopt the following representation of ν,

$$\nu = \frac{2\sigma^2\beta^2}{1+\alpha}, \tag{2.45}$$

which is equivalent to the original formula with R set to 1 so that the singularity of $\alpha = 1$ is removed. This formulation also leads to additional tractability - we can write the variance as a function of ρ as follows.

Prop 2.5. The variance of Q_t, denoted by ν in Eq. (2.45), can be shown to be a function of ρ in the form of

$$\nu = \frac{4\sigma^2\beta^2 R}{(R+1)^2 + 2\rho R \pm \sqrt{(R^2-1)^2 + 4\rho^2 R^2}}. \tag{2.46}$$

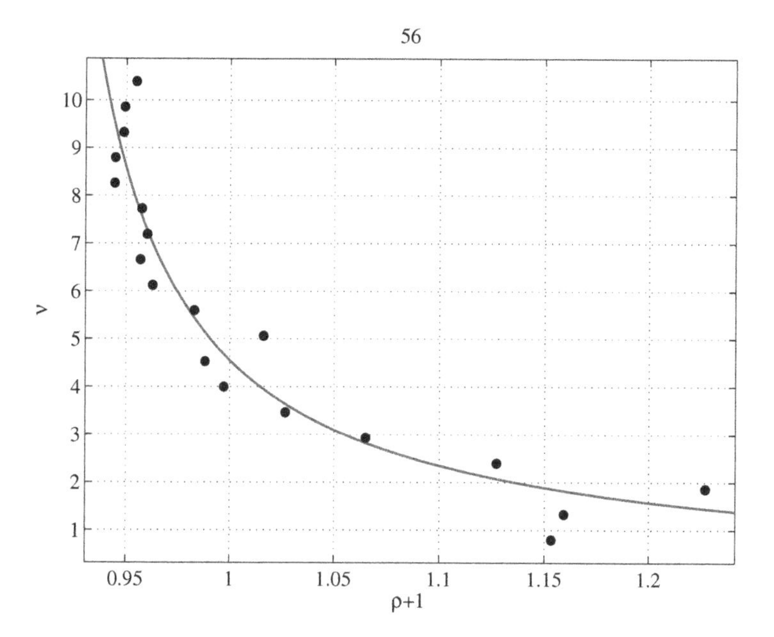

Fig. 2.5: The level of sample variance (ν) against the level of sample autocorrelation ($\rho + 1$) of the 56th portfolio daily returns of the 100 Fama-French portfolios constructed by size and book-to-market ratio. We use its historical daily returns of the period 1950-2013 and compute their rolling variances and autocorrelation with a window of 5 years. We then divide the variances into 20 bins and compute the average of the autocorrelations for each corresponding bins.

The proof of this proposition is in the appendix of this chapter.

Figure 2.6 plots the relationship between ν and $\rho + 1$ under the formula given in Proposition 2.5. Note that this is consistent with the majority of the empirical relationship shown in the 100 Fama-French portfolios like Figure 2.5 for the 56th portfolio.

2.3.3.3 *Average trade size*

The trading size of a stock is an equilibrium quantity. We may consider it as the absolute change of the shares traded by the market participants between two consecutive market clearings. Since we have assumed that the risk aversion of the market makers is constant, we define the trade size at

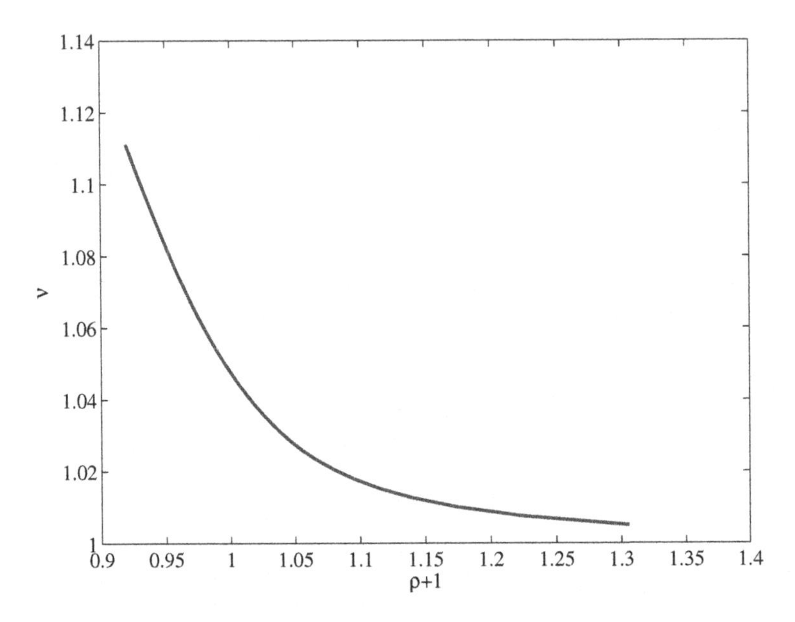

Fig. 2.6: The level of variance (ν) against the level of autocorrelation ($\rho + 1$) according to the formula in Proposition 2.5 with $\sigma = \beta = 1$.

time t by

$$V_t = w|X_t^a - X_{t-1}^a| \tag{2.47}$$

from the view of the market makers for simplicity. The market clearing condition normalizes the total shares outstanding to be 1; therefore, the trade size we describe here is actually closely related to the turnover of the stock. From Eq. (2.38), we see that trading volumes are indeed driven by the shift of the risk aversion of the marginal investors in the market. For instance, when trading cost c_λ is assumed to be zero, we have

$$V_t = \frac{w}{a}|Z_t - Z_{t-1}|, \tag{2.48}$$

that is, trade size is induced by the change of risk aversion of the marginal investors. If we define the average trade size $\bar{V}$ by the expected value of V_t,

$$\bar{V} = EV_t, \tag{2.49}$$

then we may actually obtain a closed-form solution of $\bar{V}$ by the following proposition.

Prop 2.6. In the case of zero transaction cost, the average trade size can be shown to have the following form

$$\bar{V} = \frac{2w(R-\alpha)}{a\sigma\beta\sqrt{\pi(1+\alpha)}}. \tag{2.50}$$

Combined with Eq. (2.44) and Eq. (2.45), we have

$$\frac{\sqrt{\pi\nu}a}{2\sqrt{2}w}\bar{V} + 1 = \frac{R+1}{1+\alpha}. \tag{2.51}$$

When the transaction cost is not zero, we have

$$\begin{aligned}\bar{V} &= \frac{2w(R-\alpha)}{\sqrt{\pi(\alpha+1)(a-2c_a/\beta^2\sigma^2)(a-(1-\alpha)c_a/\beta^2\sigma^2)}} \\ &= \frac{2w\sqrt{2\nu}(R-\alpha)}{\sqrt{\pi\left[a\nu(1+\alpha)-4c_a\right]\left[a\nu(1+\alpha)-2(1-\alpha)c_a\right]}},\end{aligned} \tag{2.52}$$

where $\bar{V}$ is monotonically increasing in c_a and w but decreasing in ν and α. Since ρ increases as α increases, $\bar{V}$ is decreasing in ρ, too.

The proof of this proposition is in the appendix of this chapter.

In the next section, we will use this equality to analyze how the autocorrelation of assets' excess returns are influenced by various factors in equilibrium.

2.3.4 *Model implications*

We explore here the model implications of how factors such as price-to-book ratio, return volatility, interest rate, flow of information, trade size and trading impact costs affect the autocorrelation in asset excess returns.

2.3.4.1 *Effects of risk aversion*

In our economy, if all of a sudden, due to exogenous factors, the marginal investors become significantly more risk averse than the previous period, then since changes in risk aversion will lead to actual trading volume, we should expect in Eq. (2.51) $\bar{V}$ will increase in size. Provided all other parameters fixed, a large average trade size will lead to a lower autocorrelation. Therefore, a drop in price in such situations will more likely to be followed by a price reversion in subsequent days, so that the market makers are compensated by taking on the risk in addition to the associated trading fees and beneficial impacts incurred when they accommodate the market takers' selling orders.

2.3.4.2 *Effects of private information*

However, there may be a second scenario, where it is also possible that trading activities are not primarily driven by a significant shift of risk aversion but by private information of the market takers. When such events happen, naturally the informed traders will take advantage of this information. But they will be cautious in managing their trade sizes so as to avoid being discovered by other market participants. Since once this information becomes public, all investors will follow suit, adjusting the stock price to reflect this information. Hence, the trade size should not change much. Since this trading activity does not source from the change of the risk aversion of the marginal investors, it must come from the trading cost c_a. For instance, if the private information suggests a buy on the stock, then the informed traders will most likely to continue taking on all available offers and eventually elevate the price of the security. In this case, Eq. (2.52) suggests the autocorrelation of the asset return will increase. That is, informed trading is likely to be accompanied by a higher autocorrelation.

2.3.4.3 *Effects of price-to-book ratio*

To understand the effect of the style factor on return serial correlation, we utilize the price-to-book ratio, β. Specifically, value stocks are associated with low β, whereas growth stocks are featured by high β. Equation (2.45) suggests that for a fixed level of variance, the serial correlation of excess returns increases in the level of price-to-book ratio, β. That is, stocks that are undervalued tend to have a lower autocorrelation. However, since the variance ν can be impacted by β as well, i.e. $\nu = \nu(\beta)$, the actual relationship between β and ρ can practically be more complex.

In the situation that return volatility is inversely related to the price-to-book ratio (that is when value stocks have high volatility), it is apparent that the equality suggests autocorrelation will rise along with β, i.e. value stocks will have lower autocorrelation compared to growth stocks. This situation in fact corresponds to times of economic distress, during which value stocks tend to show higher volatility than growth stocks but lower autocorrelation. The autocorrelation behavior in the most recent financial crisis validates this implication. However, if variance $\nu(\beta)$ increases in the price-to-book ratio, then the dependence of ρ on β will be influenced by the competition between $\nu(\beta)$ and β^2. When $\nu(\beta)$ increases in β at a speed that is slower than the quadratic growth rate, the term β^2 will dominate and makes ρ an increasing function in β; otherwise, ρ declines with β. Therefore, effects of price-to-book ratio need to be controlled by return volatility to deduce meaningful conclusions. Since the latter case where volatility increases in the price-to-book ratio cross-sectionally is common for non-crisis periods, the style effects can often be weak. The empirical results from Chapter 1 on the 100 Fama-French equity portfolios are consistent with the implication here.

2.3.4.4 *Effects of market capitalization*

In our model, we do not have a parameter which directly describes the market capitalization of a security. However, the following 3 features will offer some implicit insights into how the size effect takes place in our model and influences the return autocorrelation.

(1) Small caps have relatively less shares outstanding and less volume. They are often less frequently traded with larger bid/ask spreads. So on average the trade order size of a small-cap stock normalized by its shares outstanding tends to be bigger than that of large caps. Therefore, $\bar{V}$ for small caps should be bigger.

(2) Small caps have less analyst coverage, so information on small-cap stocks is less transparent and less available to the majority of in-

vestors compared with large caps. Therefore, we may assume more occurrence of effective insider trading for small caps [Lakonishok and Lee (2001)][7].

(3) Since almost all traders including retail investors, institutional investor, index funds, etc. actively trade large-cap stocks, order imbalances in large caps tend to be absorbed rather easily and quickly. Small caps trade much less frequently. It is often hard to have orders filled. Hence, the proportion of market makers w for large caps should be bigger than that of small caps.

Hence, according to our model, (1) and (3) suggest individual stocks with small market capitalization should have lower autocorrelation, whereas (2) indicates small caps will have higher autocorrelation. Hence, factors affecting the autocorrelation of the same group of stocks may have mixed implications. The three arguments so far are made only for individual stocks and they may not all applicable to pooled investments such as ETFs. For instance, some of the ETFs which track indices of small caps do not necessarily themselves experience low volume. So (1) and (3) actually cannot significantly distinguish them from ETFs tracking large-cap indices. (2) can potentially be applicable to ETFs, however, since insider information is not limited to the actual bid-ask spread or relevant trading capacity concerns. In addition, index ETFs essentially represent baskets of individual stocks, therefore, when estimating their NAVs, non-synchronous trading can be a meaningful factor to their autocorrelations as well, though this model under discussion is not equipped to contain this contribution.

2.3.5 *Empirical effects of average trade size on the autocorrelation of individual stocks*

We utilize the CRSP database and consider all stocks whose "number of trades" data entry is available.[8] We compute the daily return autocorre-

[7] Lakonishok and Lee examined insider trading activities of all NYSE, AMEX and Nasdaq stocks during the 1975-1995 period. They found that insiders seemed to be able to better predict small-cap firms.

[8] According to the database instruction, only NASDAQ stocks have such values.

lation for every three-year sub-periods between 1996 - 2013 for all tickers whose number of trading days is greater than 99% of the total trading days for the sub-period. This way, stale prices may be less of a concern. We define the average trade size for a period $[1, T]$ by averaging over the daily trading volume in shares dividing both the total number of shares outstanding and the number of trades occurred in the day, i.e.,

$$trade\ size_t = \frac{volume_t}{shares\ outstanding_t \cdot number\ of\ trades_t} \tag{2.53}$$

and

$$average\ trade\ size_{[1,T]} = \frac{1}{T}\sum_{t=1}^{T} trade\ size_t, \tag{2.54}$$

where T is the total number of days for the period $[1, T]$. Our model suggested that the average trade size should be inversely related to the autocorrelation of stock returns. We therefore attempt to validate this model implication empirically here.

For each three-year period, after computing the autocorrelation, we first rank all stocks by their average trade size, then regroup them into 6 portfolios in the ranked order. This way, the portfolios can be distinguished by the level of average trade size of the stocks. The average autocorrelation and its standard error, the number of stocks within the portfolio, the number of common trading days are all reported in Table 2.2 and Table 2.3. In particular, Table 2.2 shows the statistics for the period 1996-2004, whereas Table 2.3 shows the statistics for 2005-2013.

In the order of the average trade size, the portfolios are numbered from 1 to 6, with portfolio 1 having the smallest average trade size, whereas portfolio 6 having the largest average trade size. We see that as the average trade size increases from left to right, the average autocorrelation of the corresponding portfolio individual stock daily returns decline. For instance, in the period 2008-2010, as the portfolio number increases from 1 to 6, the corresponding average serial correlation drops from -0.0414 to -0.2342. Specifically, for Portfolio 6, the standard error is only 0.1279, leaving the average correlation almost two standard deviations away from 0.

1996-1998						
portfolio number	1	2	3	4	5	6
average trade size	0.0403	0.0891	0.1349	0.1911	0.2797	0.6067
$\bar{\rho}$	-0.0511	-0.0717	-0.1151	-0.1454	-0.22	-0.2554
standard error	0.1028	0.1301	0.1320	0.1301	0.1237	0.1235
market capitalization	2.0606	0.2805	0.1303	0.0856	0.0514	0.0266
number of stocks	449	439	437	432	435	433
number of trades	729	733	726	727	736	729
1999-2001						
portfolio number	1	2	3	4	5	6
average trade size	0.0154	0.0392	0.0649	0.1015	0.1634	0.389
$\bar{\rho}$	-0.0301	-0.0443	-0.0708	-0.1244	-0.1962	-0.2344
standard error	0.0918	0.1059	0.1080	0.1202	0.1143	0.1061
market capitalization	6.2369	0.3785	0.173	0.0867	0.0503	0.0266
number of stocks	387	386	390	391	391	394
number of trades	739	745	735	742	742	741
2002-2004						
portfolio number	1	2	3	4	5	6
average trade size	0.0053	0.013	0.0224	0.0379	0.0737	0.2559
$\bar{\rho}$	-0.0384	-0.0405	-0.0389	-0.061	-0.1329	-0.1973
standard error	0.0754	0.0854	0.0950	0.0955	0.1034	0.1000
market capitalization	4.4276	0.4374	0.2253	0.1202	0.0622	0.0313
number of stocks	415	417	416	416	415	417
number of trades	741	742	729	740	731	716

Table 2.2: The statistics of autocorrelation with average trade size for the period 1996-1998, 1999-2001, 2002-2004. For each three-year period, after computing the autocorrelation, we rank all stocks by their average trade size and regroup them into 6 portfolios in the ranked order, so that the portfolios can be distinguished by the level of average trade size of the stocks. The average autocorrelation $\bar{\rho}$ and its standard error, the number of stocks within the portfolio, the number of common trading days are reported.

2005-2007						
portfolio number	1	2	3	4	5	6
average trade size	0.0027	0.0063	0.01	0.016	0.0339	0.1142
$\bar{\rho}$	-0.0183	-0.0231	-0.0284	-0.0416	-0.0707	-0.1765
standard error	0.0552	0.0638	0.0703	0.0688	0.0982	0.1162
market capitalization	6.5332	0.6816	0.4081	0.2231	0.1183	0.0564
number of stocks	359	365	361	358	362	364
number of trades	749	743	741	744	739	731
2008-2010						
portfolio number	1	2	3	4	5	6
average trade size	0.0018	0.0044	0.0073	0.0124	0.026	0.1081
$\bar{\rho}$	-0.0414	-0.0375	-0.0533	-0.0629	-0.1036	-0.2342
standard error	0.0731	0.0749	0.0856	0.1007	0.1107	0.1279
market capitalization	4.6432	1.3504	0.3642	0.8375	0.5518	0.0807
number of stocks	380	377	374	372	371	378
number of trades	749	745	731	733	713	691
2011-2013						
portfolio number	1	2	3	4	5	6
average trade size	0.0013	0.0034	0.0058	0.0101	0.0218	0.2191
$\bar{\rho}$	-0.0339	-0.0484	-0.0576	-0.0723	-0.0865	-0.1682
standard error	0.0657	0.0691	0.0721	0.0942	0.1068	0.1343
market capitalization	10.3826	1.0239	0.5008	0.2606	0.0997	0.0406
number of stocks	332	335	329	325	330	330
number of trades	745	741	742	753	747	742

Table 2.3: The statistics of autocorrelation with average trade size for the period 2005-2007, 2008-2010, 2011-2013. For each three-year period, after computing the autocorrelation, we rank all stocks by their average trade size and regroup them into 6 portfolios in the ranked order, so that the portfolios can be distinguished by the level of average trade size of the stocks. The average autocorrelation $\bar{\rho}$ and its standard error, the number of stocks within the portfolio, the number of common trading days are reported.

In addition, we observe that the average trade size is inversely associated with market capitalization of the stocks. For example, during the period 2011-2013, as the portfolio number increases from 1 to 6, the corresponding market capitalization decreases from 10.3826 to 0.0406. Portfolios with large average trade size tend to be characterized by small cap stocks with significantly lower individual return autocorrelation average. This is consistent with our model's prediction.

2.4 Conclusion

In this chapter, we have considered two models on return autocorrelation. The first one assumes that asset returns follow a one-factor model and the factor is autocorrelated in the presence of stale prices. Results show that non-synchronous trading and factor autocorrelation both play important roles in explaining the magnitude of portfolio return autocorrelation. In particular, given the fact that many broad market equity indices and ETFs show statistically significant negative autocorrelation for the 2006-2013 period, it is clear that non-trading alone is insufficient. The assumption of autocorrelated factor is also consistent with empirical findings on Fama-French factors, as we have seen that all three factors including the market premium, small-minus-large, low-minus-high exhibit strong autocorrelations throughout the period 1926-2013, ranging from below -0.2 to above 0.4.

The model also helps to explain Figure 1.4 in Chapter 1, where we see the evolutions of the return autocorrelations of three size portfolios and that the small-cap portfolio almost always have return autocorrelation higher than the large-cap portfolio, regardless whether its autocorrelation is above 0.5 or below 0. In the situation where the return autocorrelations of all three portfolios are negative, non-synchronous trading may still be effective in justifying the amount of autocorrelation which the small-cap portfolio has over the autocorrelation of the large-cap portfolio.

Furthermore, this model provides support to understanding the discrepancies between the corporate bond ETFs and their tracking bond indices. For instance, as we have reported in Chapter 1, the daily return autocorrelation of HYG is slightly negative but the autocorrelation of its tracking index is over 0.5. Given the stale prices of corporate bonds, the non-trading model adds additional insights. Recall that the illustrated non-trading model assumes zero observed return when a portfolio constituent does not trade. However, when calculating the levels of HYG's underlying index, fund managers tend to compute the return using estimated prices for bonds that did not trade. The estimates may be based on similar securities' execution prices, movements in market rates, credit-spread changes, etc. If indeed the prices of the similar securities are used, then potentially more index autocorrelation is induced, as such estimates likely add to the cross-correlations among bonds which make up the majority of the total autocorrelation for the index.

The second model is based on the trading activities between risk-averse market makers and market takers. It helps explain how various factors influence the behavior of the excess return autocorrelation of stock returns. The model implies an inverse relationship between the autocorrelation of security returns and its average trade size but positive relation with interest rates; insider trading likely leads to positive autocorrelation. Also, high volatility is often associated with low return autocorrelation and value stocks tend to show lower autocorrelation during distressed market environments. NASDAQ stock data for the period 1996 - 2013 have been used to confirm empirically that low autocorrelation of individual stocks is often associated with trades of large sizes and often small caps tend to experience big average trade sizes.

2.5 Appendix for Chapter 2

2.5.1 *Proof of Proposition 2.1*

First, we note that $E(r_{it}^o) = \mu_i$ is trivial. To obtain the second equation in the proposition, we evaluate

$$\begin{aligned} &Var(r_{it}^o) \\ &= E(r_{it}^o)^2 - \mu_i^2 \end{aligned}$$

$$= E\left(\sum_{k=0}^{\infty}\sum_{q=0}^{\infty} X_{it}(k)X_{it}(q)r_{it-k}r_{it-q}\right) - \mu_i^2.$$

Since $X_{it}(k)X_{it}(q) = X_{it}(k)$, if $k \geq q$, and

$$\begin{aligned}
&Er_{it-k}r_{it-q} \\
&= \mu_i^2 + \beta_i^2 Ef_{t-k}f_{t-q} + Ee_{it-k}e_{it-q} \\
&= \mu_i^2 + \frac{\beta_i^2\sigma_f^2}{1-\alpha^2} + \sigma_i^2 \text{ if } k = q \\
&\text{or} \\
&= \mu_i^2 + \frac{\beta_i^2\sigma_f^2\alpha^{|k-q|}}{1-\alpha^2} \text{ if } k \neq q\ ,
\end{aligned}$$

we can get

$$\begin{aligned}
&Var(r_{it}^o) \\
&= \sum_{k=0}^{\infty}(1-\pi_i)\pi_i^k\left(\mu_i^2 + \frac{\beta_i^2\sigma_f^2}{1-\alpha^2} + \sigma_i^2\right) \\
&+ 2\sum_{k>q}(1-\pi_i)\pi_i^k\left(\mu_i^2 + \frac{\beta_i^2\sigma_f^2\alpha^{k-q}}{1-\alpha^2}\right) - \mu_i^2 \\
&= \frac{\beta_i^2\sigma_f^2}{1-\alpha^2} + \sigma_i^2 + 2\sum_{q=0}^{\infty}\mu_i^2\pi_i^{q+1} + 2\sum_{q=0}^{\infty}\frac{(1-\pi_i)\beta_i^2\sigma_f^2\alpha^{-q}(\pi_i\alpha)^{q+1}}{(1-\alpha^2)(1-\pi\alpha)} \\
&= \frac{2\pi_i}{1-\pi_i}\mu_i^2 + \frac{\beta_i^2\sigma_f^2}{1-\alpha^2} + \sigma_i^2 + \frac{2\beta_i^2\sigma_f^2\alpha\pi_i}{(1-\alpha^2)(1-\pi_i\alpha)} \\
&= \frac{2\pi_i}{1-\pi_i}\mu_i^2 + \beta_i^2\sigma_f^2 + \sigma_i^2 + \frac{2\beta_i^2\sigma_f^2\alpha\pi_i}{(1-\alpha^2)(1-\pi_i\alpha)} + \frac{\alpha^2\beta_i^2\sigma_f^2}{1-\alpha^2}.
\end{aligned}$$

For the auto-covariance of realized returns, r_{it}^o, we may compute

$$\begin{aligned}
&Cov(r_{it}^o, r_{it+1}^o) \\
&= E\left(\sum_{k=0}^{\infty}\sum_{q=0}^{\infty} X_{it}(k)X_{it+1}(q)r_{it-k}r_{it+1-q}\right) - \mu_i^2.
\end{aligned}$$

Since

$$X_{it}(k)X_{it+1}(q) = 0 \tag{2.55}$$

for $q > 0$, we must have $q = 0$. Hence,

$$Cov(r_{it}^o, r_{it+1}^o)$$

$$= \sum_{k=0}^{\infty}(1-\pi_i)^2\pi_i^k E r_{it-k} r_{it+1} - \mu_i^2$$
$$= \sum_{k=0}^{\infty}(1-\pi_i)^2\pi_i^k \left(\mu_i^2 + \beta_i^2 E f_{t-k} f_{t+1}\right) - \mu_i^2$$
$$= -\pi_i\mu_i^2 + (1-\pi_i)^2\beta_i^2 \sum_{k=0}^{\infty} \frac{\sigma_f^2\alpha^{1+k}\pi_i^k}{1-\alpha^2}$$
$$= -\pi_i\mu_i^2 + \frac{(1-\pi_i)^2\beta_i^2\sigma_f^2\alpha}{(1-\alpha^2)(1-\alpha\pi_i)}.$$

To get cross-correlation,

$Cov(r_{it}^o, r_{jt+1}^o)$

$$= E\left(\sum_{k=0}^{\infty}\sum_{q=0}^{\infty} X_{it}(k)X_{jt+1}(q) r_{it-k} r_{jt+1-q}\right) - \mu_i\mu_j$$
$$= \left(\sum_{k=0}^{\infty}\sum_{q=0}^{\infty}(1-\pi_i)\pi_i^k(1-\pi_j)\pi_j^q E r_{it-k} r_{jt+1-q}\right) - \mu_i\mu_j$$
$$= \beta_i\beta_j(1-\pi_i)(1-\pi_j)\left\{\sum_{k=0}^{\infty}\pi_i^k\pi_j^{k+1}\frac{\sigma_f^2}{1-\alpha^2}\right.$$
$$\left. + \sum_{q>k+1}^{\infty}\pi_i^k\pi_j^q\frac{\sigma_f^2}{1-\alpha}\alpha^{q-k-1} + \sum_{k>q-1}^{\infty}\pi_i^k\pi_j^q\frac{\sigma_f^2}{1-\alpha}\alpha^{k+1-q}\right\}$$
$$= \frac{\beta_i\beta_j(1-\pi_i)(1-\pi_j)\pi_j}{1-\pi_i\pi_j}\frac{\sigma_f^2}{1-\alpha^2}$$
$$+ \beta_i\beta_j(1-\pi_i)(1-\pi_j)\left\{\sum_{k=0}^{\infty}\pi_i^k\frac{\sigma_f^2}{1-\alpha}\alpha^{-k-1}\frac{(\pi_j\alpha)^{k+2}}{1-\pi_j\alpha}\right.$$
$$\left. + \sum_{q=0}^{\infty}\pi_j^q\frac{\sigma_f^2}{1-\alpha}\alpha^{-q+1}\frac{(\pi_i\alpha)^q}{1-\pi_i\alpha}\right\}$$
$$= \frac{\beta_i\beta_j(1-\pi_i)(1-\pi_j)\pi_j}{1-\pi_i\pi_j}\frac{\sigma_f^2}{1-\alpha^2}$$
$$+ \frac{\beta_i\beta_j(1-\pi_i)(1-\pi_j)}{1-\pi_i\pi_j}\left\{\frac{\sigma_f^2\alpha\pi_j^2}{(1-\alpha^2)(1-\pi_j\alpha)} + \frac{\sigma_f^2\alpha}{(1-\alpha^2)(1-\pi_i\alpha)}\right\}$$
$$= \frac{\beta_i\beta_j(1-\pi_i)(1-\pi_j)\pi_j}{1-\pi_i\pi_j}\sigma_f^2 + \frac{\beta_i\beta_j(1-\pi_i)(1-\pi_j)\pi_j}{1-\pi_i\pi_j}\frac{\sigma_f^2\alpha^2}{1-\alpha^2}$$

$$+\frac{\beta_i\beta_j(1-\pi_i)(1-\pi_j)}{1-\pi_i\pi_j}\left\{\frac{\sigma_f^2\alpha\pi_j^2}{(1-\alpha^2)(1-\pi_j\alpha)}+\frac{\sigma_f^2\alpha}{(1-\alpha^2)(1-\pi_i\alpha)}\right\}.$$

2.5.2 *Proof of Proposition 2.2*

By the law of large numbers, when N is sufficient large, the portfolio virtual return

$$r_t = \mu + \beta f_t + \frac{\sum_{i=1}^N e_{it}}{N} \tag{2.56}$$

approaches

$$r_t = \mu + \beta f_t. \tag{2.57}$$

Then, it is straightforward that

$$Var(r_t) = \beta^2\frac{\sigma_f^2}{1-\alpha^2}. \tag{2.58}$$

To compute $Var(r_t^o)$, we follow the next steps

$$\begin{aligned}
&Var(r_t^o)\\
&=\frac{1}{N^2}\left\{\sum_{i=1}^N \frac{2\pi}{1-\pi}\mu_i^2+\beta_i^2\sigma_f^2+\sigma_i^2+\frac{2\beta_i^2\sigma_f^2\alpha\pi}{(1-\alpha^2)(1-\pi\alpha)}+\frac{\alpha^2\beta_i^2\sigma_f^2}{1-\alpha^2}\right.\\
&\left.+\sum_{j\neq i}\left[\frac{\beta_i\beta_j(1-\pi)}{1+\pi}\cdot\frac{\sigma_f^2}{1-\alpha^2}\left(1+\frac{2\pi\alpha}{1-\pi\alpha}\right)\right]\right\}\\
&=\beta^2\frac{1-\pi}{1+\pi}\cdot\frac{\sigma_f^2}{1-\alpha^2}\cdot\frac{1+\pi\alpha}{1-\pi\alpha}+O(1/N).
\end{aligned}$$

Now,

$$\begin{aligned}
&Cov(r_t^o, r_{t+1}^o)\\
&=\frac{1}{N^2}\left\{\sum_{i=1}^N\left[-\pi\mu_i^2+\frac{(1-\pi)^2\beta_i^2\sigma_f^2\alpha}{(1-\alpha^2)(1-\alpha\pi)}\right]\right.\\
&\left.+\sum_{j\neq i}\left[\frac{\beta_i\beta_j(1-\pi)}{1+\pi}\cdot\frac{\sigma_f^2}{1-\alpha^2}\left(\pi+\frac{2\alpha}{1-\pi\alpha}\right)\right]\right\}
\end{aligned}$$

$$= \beta^2 \frac{1-\pi}{1+\pi} \frac{\sigma_f^2}{1-\alpha^2} \left(\pi + \frac{\alpha}{1-\pi\alpha} + \frac{\alpha\pi^2}{1-\pi\alpha} \right) + O(1/N).$$

Therefore, as $N \to \infty$, we have

$$\begin{aligned} \rho_\alpha &= \frac{\pi+\alpha}{1+\pi\alpha} \\ &= \pi + \alpha - \frac{\pi\alpha(\pi+\alpha)}{1+\pi\alpha}. \end{aligned}$$

2.5.3 *Proof of Proposition 2.4*

Since

$$Q_{t+1} = (1-\alpha)\beta\bar{B} + (\alpha - R)P_t + \beta u_t, \tag{2.59}$$

the variance of Q_{t+1} is

$$\begin{aligned} \nu &= Var\left[-\beta r\bar{B} + (\alpha - R)\beta\tilde{B}_t + \beta u_{t+1}\right] \\ &= \beta^2(\alpha - R)^2 Var\tilde{B}_t + \beta^2 Var\ u_{t+1} \\ &= \beta^2\left[(\alpha - R)^2\frac{\sigma^2}{1-\alpha^2} + \sigma^2\right] \\ &= \frac{\sigma^2\beta^2(1+R^2-2\alpha R)}{1-\alpha^2}. \end{aligned} \tag{2.60}$$

The first lag auto-covariance $\gamma(1)$ can be shown to be

$$\begin{aligned} \gamma(1) &= Cov(Q_{t+1}, Q_t) \\ &= E\left[(\alpha - R)\beta B_{t+1} + \beta u_{t+2}\right]\left[(\alpha - R)\beta B_t + \beta u_{t+1}\right] \\ &= \beta^2 E\left[(\alpha - R)^2 B_{t+1}B_t + (\alpha - R)B_{t+1}u_{t+1}\right] \\ &= \beta^2\left[(\alpha - R)^2\frac{\sigma^2\alpha}{1-\alpha^2} + (\alpha - R)\sigma^2\right] \\ &= \frac{\sigma^2\beta^2(\alpha - R)}{1-\alpha^2}\left[\alpha(\alpha - R) + 1 - \alpha^2\right] \\ &= \frac{\sigma^2\beta^2(R-\alpha)(\alpha R - 1)}{1-\alpha^2}. \end{aligned} \tag{2.61}$$

Therefore, autocorrelation is of the form

$$\rho = \frac{(R-\alpha)(\alpha R - 1)}{1+R^2-2\alpha R}. \tag{2.62}$$

Since we have assumed that $|\alpha| < 1$,

$$1 + R^2 - 2\alpha^2 R = (1 - R)^2 + 2R(1 - \alpha) > 0. \tag{2.63}$$

Therefore,

$$\rho > 0, \text{ when } \frac{1}{R} < \alpha < 1 \tag{2.64}$$

and

$$\rho < 0, \text{ when } \alpha < \frac{1}{R}. \tag{2.65}$$

Since

$$\frac{\partial \rho}{\partial R} = \frac{(1 - \alpha^2)(R^2 - 1)}{(1 + R^2 - 2\alpha R)^2} > 0, \tag{2.66}$$

we have ρ is monotonic increasing in R.

Next, we will show that ρ is also increasing in α. Simple calculation gives

$$\frac{\partial \rho}{\partial \alpha} = \frac{1 + R^4 + 2\alpha^2 R^2 - 2\alpha(R^3 + R)}{(1 + R^2 - 2\alpha R)^2}. \tag{2.67}$$

We only need to show that the numerator of this partial derivative is always positive. For which, we compute

$$\begin{aligned} &\frac{\partial}{\partial \alpha}\left[1 + R^4 + 2\alpha^2 R^2 - 2\alpha(R^3 + R)\right] \\ &= 4\alpha R^2 - 2(R^3 + R) \\ &= -2R(R^2 + 1 - 2\alpha R) < 0. \end{aligned} \tag{2.68}$$

Hence, the minimum of the numerator will be reached at $\alpha = 1$, which in this case reduces to

$$\begin{aligned} &1 + R^4 + 2R^2 - 2(R^3 + R) \\ &= (1 + R^2)^2 - 2R(R^2 + 1) \\ &= (R^2 + 1)(R - 1)^2 > 0. \end{aligned} \tag{2.69}$$

We are done.

2.5.4 *Proof of Proposition 2.5*

Solving the quadratic Eq. (2.44) for α, we find

$$\alpha = \frac{1 + 2\rho R + R^2 \pm \sqrt{R^4 + (4\rho^2 - 2)R^2 + 1}}{2R}. \tag{2.70}$$

Then, $\alpha + 1$ can be written as

$$\alpha + 1 = \frac{(1 + R)^2 + 2\rho R \pm \sqrt{(R^2 + 1)^2 + 4\rho^2 R^2}}{2R}. \tag{2.71}$$

Therefore,

$$\nu = \frac{4\sigma^2 \beta^2 R}{(R + 1)^2 + 2\rho R \pm \sqrt{(R^2 - 1)^2 + 4\rho^2 R^2}}. \tag{2.72}$$

2.5.5 *Proof of Proposition 2.3*

Using the dynamics of P_t and B_t, we get

$$\begin{aligned} W_{t+1} &= W_t(1 + r) + X_t \left[\beta B_{t+1} - \beta(1 + r)B_t\right] \\ &= W_t(1 + r) + X_t \left[\beta(\bar{B} + \tilde{B}_{t+1}) - \beta(1 + r)(\bar{B} + \tilde{B}_t)\right] \\ &= W_t(1 + r) + X_t\beta \left[\bar{B} + \alpha\tilde{B}_t + u_{t+1} - (1 + r)(\bar{B} + \tilde{B}_t)\right] \\ &= W_t(1 + r) + X_t\beta \left[(-r)\bar{B} + (\alpha - 1 - r)\tilde{B}_t + u_{t+1}\right]. \end{aligned} \tag{2.73}$$

Since u_t's are independent, identically distributed normal random variables with mean 0 and variance σ^2, we have $W_{t+1}|\mathcal{F}_t$ as a normal random variable with

$$E_t W_{t+1} = W_t(1 + r) + X_t\beta \left[-r\bar{B} + (\alpha - 1 - r)\tilde{B}_t\right] \tag{2.74}$$

and

$$Var_t W_{t+1} = \beta^2 \sigma^2 X_t^2. \tag{2.75}$$

We may represent Eq. (2.24) by

$$\max_{X_t} A_1 + A_2 X_t - \frac{A_3}{2} X_t^2 + \frac{A_4}{2}(X_t - X_{t-1})^2, \tag{2.76}$$

where

$$A_1 = W_t(1+r) \tag{2.77}$$

$$A_2 = \beta\left[(-r)\bar{B} + (\alpha - 1 - r)\tilde{B}_t\right] \tag{2.78}$$

$$A_3 = \lambda\beta^2\sigma^2 \tag{2.79}$$

$$A_4 = s(\lambda)c. \tag{2.80}$$

Assume A_3 is always positive, the objective function is quadratic and concave downward, the first order condition gives the optimal solution with

$$X_t = \frac{A_2 - A_4 X_{t-1}}{A_3 - A_4}. \tag{2.81}$$

After substitution, we get

$$X_t^\lambda = \frac{(1-R)\beta\bar{B} + (\alpha - R)\beta\tilde{B}_t - s(\lambda)cX_{t-1}^\lambda}{\lambda\beta^2\sigma^2 - s(\lambda)c}. \tag{2.82}$$

More explicitly, for the liquidity traders,

$$X_t^b = \frac{(1-R)\beta\bar{B} + (\alpha - R)\beta\tilde{B}_t + cX_{t-1}^b}{b_t\beta^2\sigma^2 + c} \tag{2.83}$$

and for the market makers

$$X_t^a = \frac{(1-R)\beta\bar{B} + (\alpha - R)\beta\tilde{B}_t - cX_{t-1}^a}{a\beta^2\sigma^2 - c}. \tag{2.84}$$

We may present this recurrence relation as

$$-r\beta\bar{B} + \beta(\alpha - R)\tilde{B}_t - (a\beta^2\sigma^2 - c)X_t^a = cX_{t-1}^a \tag{2.85}$$

and, for simplicity, we rewrite this equality as

$$A + BY_t + CX_t = DX_{t-1}, \tag{2.86}$$

where

$$\begin{aligned} X_t &= X_t^a \\ A &= -r\beta\bar{B} \\ B &= \beta(\alpha - R) \\ C &= -(a\beta^2\sigma^2 - c) \\ D &= c \end{aligned} \tag{2.87}$$

and

$$Y_t = \tilde{B}_t \tag{2.88}$$

with

$$Y_t = \alpha Y_{t-1} + u_t. \tag{2.89}$$

It follows simply that

$$\begin{aligned} Y_t &= \alpha Y_{t-1} + u_t \\ &= \alpha(\alpha Y_{t-1} + u_{t-1}) + u_t \\ &= \alpha^2 Y_{t-1} + \alpha u_{t-1} + u_t \\ &= \alpha^2(\alpha Y_{t-2} + u_{t-2}) + \alpha u_{t-1} + u_t \\ &= \alpha^2 Y_{t-2} + \alpha^2 u_{t-2} + \alpha u_{t-1} + u_t \\ &\dots \\ &= \alpha^t Y_0 + \alpha^{t-1} u_1 + \dots + \alpha u_{t-1} + u_t \\ &= \sum_{j=1}^{t} \alpha^{t-j} u_j, \end{aligned} \tag{2.90}$$

where we have assumed $Y_0 = 0$.

Now, since

$$\frac{AD}{C} + \frac{BD}{C} Y_{t-1} + DX_{t-1} = \frac{D^2}{C} X_{t-2}, \tag{2.91}$$

we add this equation to Eq. (2.86) and get

$$A\left[1 + \frac{D}{C}\right] + B\left[Y_t + \frac{D}{C} Y_{t-1}\right] + CX_t = \frac{D^2}{C} X_{t-2}. \tag{2.92}$$

Similarly, we add this to the $\frac{D^2}{C^2}$ multiple of

$$A + BY_{t-2} + CX_{t-2} = DX_{t-3} \tag{2.93}$$

and get

$$A\left[1 + \frac{D}{C} + \frac{D^2}{C^2}\right] + B\left[Y_t + \frac{D}{C} Y_{t-1} + \frac{D^2}{C^2} Y_{t-2}\right] + CX_t = \frac{D^3}{C^2} X_{t-3}. \tag{2.94}$$

Therefore, we eventually get

$$A\left[1 + \frac{D}{C} + .. + \left(\frac{D}{C}\right)^{t-1}\right] + B\left[Y_t + \frac{D}{C} Y_{t-1} + \dots + \left(\frac{D}{C}\right)^{t-1} Y_1\right] + CX_t \tag{2.95}$$

$$= \frac{D^t}{C^{t-1}} X_0$$

$$= 0,$$

where we have assumed that $X_0 = 0$.

Hence, by rearranging this equation, we get

$$\begin{aligned} X_t &= -\frac{A}{C}\left[1 + \frac{D}{C} + .. + \left(\frac{D}{C}\right)^{t-1}\right] - \frac{B}{C}\left[Y_t + \frac{D}{C}Y_{t-1} + ... + \left(\frac{D}{C}\right)^{t-1} Y_1\right] \\ &= \frac{A}{D-C}\left[1 - \left(\frac{D}{C}\right)^{t}\right] - \frac{B}{C}\sum_{j=1}^{t}\left(\frac{D}{C}\right)^{t-j} Y_j. \end{aligned} \tag{2.96}$$

Substituting back to the original notions, we obtain

$$\begin{aligned} X_t^a &= \frac{-r\bar{B}}{a\beta\sigma^2}\left[1 - \left(\frac{1}{1 - a\beta^2\sigma^2/c}\right)^{t}\right] \\ &- \frac{\beta(\alpha - R)}{c - a\beta^2\sigma^2}\sum_{j=1}^{t}\left(\frac{1}{1 - a\beta^2\sigma^2/c}\right)^{t-j}\tilde{B}_j. \end{aligned} \tag{2.97}$$

2.5.6 *Proof of Proposition 2.6*

Since the trading volume is defined by

$$V_t = w|X_t^a - X_{t-1}^a|, \tag{2.98}$$

its average is

$$\begin{aligned} \bar{V} &= E\ V_t \\ &= wE\left|\frac{(\alpha - 1 - r)(\tilde{B}_t - \tilde{B}_{t-1})}{a\sigma^2\beta}\right| \\ &= \frac{w|\alpha - 1 - r|}{a\sigma^2\beta}E\ |\tilde{B}_t - \tilde{B}_{t-1}|, \end{aligned} \tag{2.99}$$

with cost c is assumed to be zero.

Recall, for a normal random variable, Z, with mean 0 and variance σ_z^2,

$$E|Z| = 2\frac{1}{\sqrt{2\pi\sigma_z^2}}\int_0^\infty ze^{-\frac{z^2}{2\sigma^2}}dz. \tag{2.100}$$

By setting $y = \frac{z^2}{2\sigma_z^2}$, we have $dy = \frac{z}{\sigma_z^2}dz$. Thus

$$\begin{aligned} E|Z| &= \frac{2\sigma_z}{\sqrt{2\pi}} \int_0^\infty e^{-y} dy \\ &= \sigma_z \sqrt{\frac{2}{\pi}}. \end{aligned} \tag{2.101}$$

Since the innovations u_t are assumed to be iid normal, $\tilde{B}_t - \tilde{B}_{t-1}$ is a normal random variable. Its mean is evidently 0 and its variance is

$$\begin{aligned} Var[\tilde{B}_t - \tilde{B}_{t-1}] &= 2Var\tilde{B}_t - 2Cov(\tilde{B}_{t-1}, \tilde{B}_t) \\ &= 2\frac{\sigma^2}{1-\alpha^2} - 2\frac{\sigma^2\alpha}{1-\alpha^2} \\ &= 2\frac{\sigma^2}{1+\alpha}, \end{aligned} \tag{2.102}$$

by the stationary of $\tilde{B}_t$. Therefore,

$$\begin{aligned} \bar{V} &= E\ V_t \\ &= \frac{2w(R-\alpha)}{a\sigma\beta\sqrt{\pi(1+\alpha)}}. \end{aligned} \tag{2.103}$$

Now, suppose the trading cost is not zero. We will adopt the notations as before,

$$\begin{aligned} X_t &= X_t^a \\ A &= -r\beta\bar{B} \\ B &= \beta(\alpha - R) \\ C &= -(a\beta^2\sigma^2 - c) \\ D &= c \end{aligned}$$

and

$$Y_t = \tilde{B}_t$$

with

$$Y_t = \alpha Y_{t-1} + u_t.$$

Since

$$X_t = -\frac{A}{C}\left[1 + \frac{D}{C} + .. + \left(\frac{D}{C}\right)^{t-1}\right] - \frac{B}{C}\left[Y_t + \frac{D}{C}Y_{t-1} + ... + \left(\frac{D}{C}\right)^{t-1} Y_1\right]$$

$$= \frac{A}{D-C}\left[1-\left(\frac{D}{C}\right)^{t}\right]-\frac{B}{C}\sum_{j=1}^{t}\left(\frac{D}{C}\right)^{t-j}Y_j,$$

we may write

$$X_t - X_{t-1} = -\frac{A}{C}\left(\frac{D}{C}\right)^{t-1} - \frac{B}{C}\left[Y_t + \left(\frac{D}{C}-1\right)Y_{t-1} + \left(\frac{D}{C}-1\right)\frac{D}{C}Y_{t-2} + \ldots + \left(\frac{D}{C}-1\right)\left(\frac{D}{C}\right)^{t-2}Y_1\right]. \tag{2.104}$$

As seen in Eq. (2.90), Y_t is a linear combination of $u_1, \ldots, u_t$, therefore, $X_t - X_{t-1}$ will also be a linear combination of $u_1, \ldots, u_t$ and is normally distributed. In particular, the coefficient of u_j, denoted by c_j will be of the following form

$$c_j = -\frac{B}{C}\left[\frac{C(\alpha-1)}{C\alpha-D}\alpha^{t-j} - \frac{D-C}{C\alpha-D}\left(\frac{D}{C}\right)^{t-j}\right]. \tag{2.105}$$

Thus, if we denote

$$\mu_v = E(X_t - X_{t-1}) \tag{2.106}$$

and

$$\sigma_v^2 = Var(X_t - X_{t-1}), \tag{2.107}$$

then we can show that

$$\mu_v = -\frac{A}{C}\left(\frac{D}{C}\right)^{t-1} \tag{2.108}$$

and

$$\sigma_v^2 = \sigma^2\sum_{j=1}^{t}c_j^2. \tag{2.109}$$

Hence, if we denote $Z = X_t - X_{t-1}$, then $Z \sim N(\mu_v, \sigma_v^2)$ and

$$E|Z| = \sigma_v\sqrt{\frac{2}{\pi}}\exp\left(-\frac{\mu_v}{2\sigma_v^2}\right) + \mu_v\left[1-2\Phi\left(-\frac{\mu_v}{\sigma_v}\right)\right]. \tag{2.110}$$

Moreover, its partial derivatives with respect to μ_v and σ_v can be computed as

$$\frac{\partial E|Z|}{\partial \mu_v} = 1 - 2\Phi\left(-\frac{\mu_v}{\sigma_v}\right) \tag{2.111}$$

and

$$\frac{\partial E|Z|}{\partial \sigma_v} = \sqrt{\frac{2}{\pi}} \exp\left(-\frac{\mu_v}{2\sigma_v^2}\right). \tag{2.112}$$

Hence, it is increasing in σ and $|\mu_v|$.

Now, if we assume that the series do not stop at time 0 but can go back to negative infinity and that $\left|\frac{D}{C}\right| < 1$ (or $\lambda\beta^2\sigma^2 - 2c > 0$), then Z will be stationary with

$$\mu_v = 0 \tag{2.113}$$

and

$$\begin{aligned}
\sigma_v^2 &= \left(\frac{\sigma B}{C^2\alpha - CD}\right)^2 \left[\frac{C^2(\alpha-1)^2}{1-\alpha^2} + \frac{(D-C)^2}{1-\left(\frac{D}{C}\right)^2} - 2\frac{C(\alpha-1)(D-C)}{1-\frac{\alpha D}{C}}\right] \\
&= \left(\frac{\sigma B}{C\alpha - D}\right)^2 \left[\frac{(\alpha-1)^2}{1-\alpha^2} + \frac{(C-D)^2}{C^2-D^2} + 2\frac{(\alpha-1)(C-D)}{C-\alpha D}\right] \\
&= \left(\frac{\sigma B}{C\alpha - D}\right)^2 \left[\frac{1-\alpha}{1+\alpha} + \frac{C-D}{C+D} + 2\frac{(\alpha-1)(C-D)}{C-\alpha D}\right] \\
&= \left(\frac{\sigma B}{C\alpha - D}\right)^2 \frac{2(\alpha C - D)^2}{(\alpha+1)(C+D)(C-\alpha D)} \\
&= \frac{2\sigma^2 B^2}{(\alpha+1)(C+D)(C-\alpha D)}.
\end{aligned} \tag{2.114}$$

Hence, rewriting this using the original notions, we get

$$\sigma_v^2 = \frac{2\sigma^2\beta^2(R-\alpha)^2}{(\alpha+1)(a\beta^2\sigma^2 - 2c)(a\beta^2\sigma^2 - (1-\alpha)c)}. \tag{2.115}$$

Therefore,

$$\begin{aligned}
E|V| &= wE|Z| \\
&= \frac{2w\beta\sigma(R-\alpha)}{\sqrt{\pi(\alpha+1)(a\beta^2\sigma^2 - 2c)(a\beta^2\sigma^2 - (1-\alpha)c)}} \\
&= \frac{2w(R-\alpha)}{\sqrt{\pi(\alpha+1)(a - \frac{2c}{\beta^2\sigma^2})(a - \frac{(1-\alpha)c}{\beta^2\sigma^2})}} \\
&= \frac{2w\sqrt{2\nu}(R-\alpha)}{\sqrt{\pi\left[a\nu(1+\alpha) - 4c\right]\left[a\nu(1+\alpha) - 2(1-\alpha)c\right]}}.
\end{aligned} \tag{2.116}$$

Now, to see the monotonicity of $\bar{V}$ with parameters c, ν and α, we compute the corresponding partial derivatives:

$$\frac{\partial \bar{V}}{\partial c} = -\frac{2\sqrt{2\nu}w(R-\alpha)\left[a(\alpha-3)(\alpha+1)\nu+8(1-\alpha)c\right]}{\sqrt{\pi}\left[(a(\alpha+1)\nu-4c)(a(\alpha+1)\nu+2(\alpha-1)c)\right]^{3/2}}. \tag{2.117}$$

Since

$$a\nu(1+\alpha) > 4c, \tag{2.118}$$

we have

$$\begin{aligned} & a(\alpha-3)(\alpha+1)\nu+8c(1-\alpha) \\ < & 4c(\alpha-2)+8c(1-\alpha) \\ = & -4c(1+\alpha), \end{aligned}$$

where $\alpha - 3 < 0$. Therefore, together with the negative sign before the quotient, we have $\frac{\partial \bar{V}}{\partial c} > 0$. That is large trade sizes tend to be associated with high costs.

The partial derivative of $\bar{V}$ with respect to ν can be computed as

$$\frac{\partial \bar{V}}{\partial \nu} = -\frac{\sqrt{2}w(R-\alpha)\left[a^2(\alpha+1)^2\nu^2-8(1-\alpha)c^2\right]}{\sqrt{\pi\nu}\left[(a(\alpha+1)\nu-4c)(a(\alpha+1)\nu+2(\alpha-1)c)\right]^{3/2}}. \tag{2.119}$$

Apply the inequality in Eq. (2.118) and get easily that $\frac{\partial \bar{V}}{\partial \nu} < 0$, since $1-\alpha > 0$; therefore, small trade sizes tend to be accompanied by high volatility environments.

The partial derivative of $\bar{V}$ with respect to α can be computed as

$$\frac{\partial \bar{V}}{\partial \alpha} = -\frac{2\sqrt{2\nu}w(R-\alpha)\tau(\alpha)}{\sqrt{\pi}\left[(a(\alpha+1)\nu-4c)(a(\alpha+1)\nu+2(\alpha-1)c)\right]^{3/2}}, \tag{2.120}$$

where

$$\tau(\alpha) = a^2(\alpha+1)(R+1)\nu^2+2ac\nu(\alpha R-\alpha-R-3)-4c^2(\alpha+R-2). \tag{2.121}$$

To show that $\frac{\partial \bar{V}}{\partial \alpha} < 0$, it is sufficient that we show $\tau(\alpha) > 0$.

Since

$$\tau(\alpha) > 4ac\nu(R+1)+2ac\nu(\alpha R-\alpha-3)-4c^2(\alpha+R-2)$$

and

$$
\begin{aligned}
&4ac\nu(R+1)+2ac\nu(\alpha R-\alpha-3)-4c^2(\alpha+R-2)\\
&=2ac\nu(R-1)(1+\alpha)-4c^2(\alpha+R-2),
\end{aligned}
$$

we must have

$$\tau(\alpha) > 2ac\nu(R-1)(1+\alpha)-4c^2(\alpha+R-2).$$

But

$$2ac\nu(R-1)(1+\alpha)-4c^2(\alpha+R-2) > 4c^2(2R-2-\alpha-R+2),$$

therefore,

$$\tau(\alpha) > 4c^2(R-\alpha).$$

We are done.

This says that $\bar{V}$ and α are inversely related. But ρ is increasing in α, thus, we must have $\bar{V}$ is inversely related to ρ. That is bigger trade sizes tend to be associated with lower autocorrelation.

Chapter 3

Discrete Sampling of Variances

3.1 Introduction

There are quite a number of ways to trade realized volatility using derivatives [Carr *et al.* (1998)][Hull (2009)]. One traditional method is to trade a delta-hedged European option. For instance, suppose an investor sells a European option at time S which matures at T for a value, $V(F_S, S; \sigma)$ computed using the Black model, and hedges it dynamically with $\frac{\partial V(F_t,t;\sigma)}{\partial F}$ futures contracts over $[S, T]$. Then, the terminal value of her trade can be calculated using Ito's formula as,

$$\begin{aligned} V(F_T, T; \sigma) = V(F_S, S; \sigma)e^{r(T-S)} &+ \int_S^T e^{r(T-t)} \frac{\partial V(F_t, t; \sigma)}{\partial F} dF_t \qquad (3.1) \\ &+ \int_S^T e^{r(T-t)} \left[-rV(F_t, t; \sigma) + \frac{\partial V(\Gamma_t, t; \sigma)}{\partial t} \right] dt \\ &+ \int_S^T e^{r(T-t)} \frac{\partial^2 V(F_t, t; \sigma)}{\partial F^2} \frac{F_t^2}{2} \sigma_t^2 dt, \end{aligned}$$

where F_t follows $dF_t = F_t \sigma_t dt$, since $V(F, t; \sigma)$ satisfies the Black partial differential equation

$$-rV(F, t; \sigma) + \frac{\partial V(F, t; \sigma)}{\partial t} = -\frac{\sigma^2 F^2}{2} \frac{\partial^2 V(F, t; \sigma)}{\partial F^2}, \qquad (3.2)$$

with

$$V(F, T; \sigma) = f(F), \qquad (3.3)$$

where $f(F)$ is the final payoff.

To see why this method relates to volatility trading. We substitute Eq. (3.2) into Eq. (3.1) and get the next expression,

$$f(F_T) + \int_S^T e^{r(T-t)} \frac{F_t^2}{2} \frac{\partial^2 V(F_t, t; \sigma)}{\partial F^2} (\sigma^2 - \sigma_t^2) dt \tag{3.4}$$
$$= V(F_T, T; \sigma) e^{r(T-S)} + \int_S^T e^{r(T-t)} \frac{\partial^2 V(F_t, t; \sigma)}{\partial F^2} dF_t.$$

The right-hand side of this equation is the terminal value of a static investment in the option contract, initially at the value $V(F_S, S; \sigma)$, and a dynamic position of holding $\frac{\partial V(F_t, t; \sigma)}{\partial F}$ futures contracts over (S, T). Therefore, the profit and loss of this trade for the period from S to T,

$$P\&L = \int_S^T e^{r(T-t)} \frac{F_t^2}{2} \frac{\partial^2 V(F_t, t; \sigma)}{\partial F^2} (\sigma^2 - \sigma_t^2) dt, \tag{3.5}$$

can be represented as the difference between the implied variance, σ^2, and the realized variance, σ_t^2, weighted by half the dollar gamma. If the implied variance is chosen initially to be the same as the realized variance, then we make neither a profit nor a loss. If $\sigma^2 > \sigma_t^2$, that is, the buyer of the option overestimates the realized variance over (S, T), then, as a seller, we collect a profit at T. For contracts like variance swaps which have constant gamma, the P&L will then become exactly the difference between the implied variance and the realized variance over the period of the trade.

In reality, however, option hedging cannot be done continuously; therefore, the discretization of the continuous time stochastic process for asset price dynamics impacts the performance of such strategies. For instance, the way in which σ_t^2 is sampled depends on the horizon at which the holder of the derivative rebalances her hedging portfolio. If the variances sampled using asset returns at different horizons for the same sample period vary quite a bit, then, instead of taking the advantage of the difference between the realized variance and the implied variance, we can also profit from the difference of the realized variances discretely sampled at different horizons. To do so, the trader can utilize her knowledge on the previous option strategy; she can go long on the option hedged at one frequency and short the same option hedged at the other frequency. Initially, this may seem rather tedious, but, interestingly, the net of these positions eliminates the exposure in options, leaving dynamic trades solely on the underlying securities, that is, we can capture the variance difference by simply trading the underlying asset without derivatives. Provided with sufficient liquidity of the underlying securities, this method can offer more flexibility to investors by allowing them to obtain not only similar payoffs as in over-the-counter swap trades but also potentially a much more diversified pool of strategies.

In earlier chapters, we have shown that the realized variances sampled at different horizons are indeed different. For instance, in Chapter 1, we see that many ETFs exhibit statistically significant variance ratios, that is, their realized variances computed at different return horizons remain different for a considerably long period of time. In Chapter 2, we investigate the sources of this market inefficiency and discover that many factors influence the magnitudes of the variance gap. Therefore, the above strategies are likely to be profitable.

In the rest of this chapter, we continue[1] the research on the discrete sampling of variance. We illustrate a systematic strategy which exploits the gap between variances discretized at different horizons in Section 3.2. In Section 3.3, we use a discretized continuous time model and several discrete time models to investigate how serial correlation affects the performance of the strategy. Section 3.4 backtests the strategy using actual securities price series. The chapter will be concluded in Section 3.5.

3.2 A variance strategy

Suppose $f(x)$ is a sufficiently smooth function in x, then Ito's lemma applied to $f(S_t)$ asserts

$$df(S_t) = f'(S_t)dS_t + \frac{1}{2}f''(dS_t)^2. \tag{3.6}$$

We shall assume that S_t follows a generalized geometric brownian motion,

$$\frac{dS_t}{S_t} = \mu(t, S_t)dt + \sigma(t, S_t)dB_t; \tag{3.7}$$

[1]Broadie and Jain [Broadie and Jain (2008)] investigated the effect of discrete sampling and asset price jumps on fair variance and volatility swap strikes. They found that under Merton jump-diffusion models and stochastic volatility with jump models, the fair discrete variance strike converges linearly to the fair continuous variance strike with the number of sampling dates. Furthermore, for realistic contract specifications and model parameters, the effect of discrete sampling is typically small while the effect of jumps can be significant.

but for simplicity, we will denote $\mu(t, S_t)$ by μ and $\sigma(t, S_t)$ by σ. By assuming that $f(S_t) = \log(S_t)$, we have

$$\log(S_T/S_0) = \int_0^T \frac{dS_t}{S_t} - \frac{1}{2}\int_0^T \sigma^2 dt. \tag{3.8}$$

Since

$$d\log S_t = \left(\mu - \frac{\sigma^2}{2}\right) dt + \sigma dB_t, \tag{3.9}$$

the quantity $\int_0^T \sigma^2 dt$ is naturally the quadratic variation of $\log S_t$, which is approximately the realized variance.

Equation (3.8), therefore, can be interpreted as follows. Its left-hand side is a static position which pays log returns at time T. The first term on the right-hand side is a dynamic position of $\frac{1}{S_t}$ shares of the asset at each time t and the second term is the realized variance up to time T. The left-hand side does not depend on how S_t evolves, thus remains the same when we discretize the right-hand side using different frequencies. Suppose h is small, then we can have the following approximation

$$\log(S_T/S_0) \approx \sum_{j=1}^{[T/h]} \frac{S_{jh} - S_{(j-1)h}}{S_{(j-1)h}} - \frac{1}{2}\left[\log S_{jh} - \log S_{(j-1)h}\right]^2. \tag{3.10}$$

If we discretize Eq. (3.8) with different scales, h and H ($H > h$), and subtract the two equations, then we can get

$$\begin{aligned} &\sum_{j=1}^{[T/h]} \frac{S_{jh} - S_{(j-1)h}}{S_{(j-1)h}} - \sum_{j=1}^{[T/H]} \frac{S_{jH} - S_{(j-1)H}}{S_{(j-1)H}} \\ &\approx \frac{1}{2}\left(\sum_{j=1}^{[T/h]} \left[\log \frac{S_{jh}}{S_{(j-1)h}}\right]^2 - \sum_{j=1}^{[T/H]} \left[\log \frac{S_{jH}}{S_{(j-1)H}}\right]^2\right), \end{aligned} \tag{3.11}$$

where the $\log(S_T/S_0)$ is cancelled. The left-hand side is the P&L of a dynamic long position of holding a constant dollar position of the underlying stock with rebalancing frequency h and a short dynamic position of the same kind with a rebalancing frequency H. The right-hand side is exactly the difference between the realized variances computed using log-returns at different horizons. Therefore, by dynamically rebalancing the underlying security alone, without trading derivatives, we can achieve the difference of realized variances at different scales.

The choice of the form of $f(x)$ can vary. For instance, in Eq. (3.6), if we assume $f(S_t) = S_t^2$, we get

$$S_T^2 - S_0^2 = 2\int_0^T S_t dS_t + \int_0^T \sigma^2 S_t^2 dt, \tag{3.12}$$

where the last term is the quadratic variation of S_t, $[S,S]_T$. Again the left-hand side is a static position which depends on the initial and terminal values of the stock only, whereas the right-hand side is a dynamic position and the realized variance weighted by the asset price, or realized dollar variance. Figure 3.1 plots the annualized discrete sampling of the dollar variances of SPY at different horizons for the period Jan 1, 2006 - Sep 5, 2013. We see that similar to Figure 1.2, the discretized dollar variances decrease as the horizon of the price difference increases.

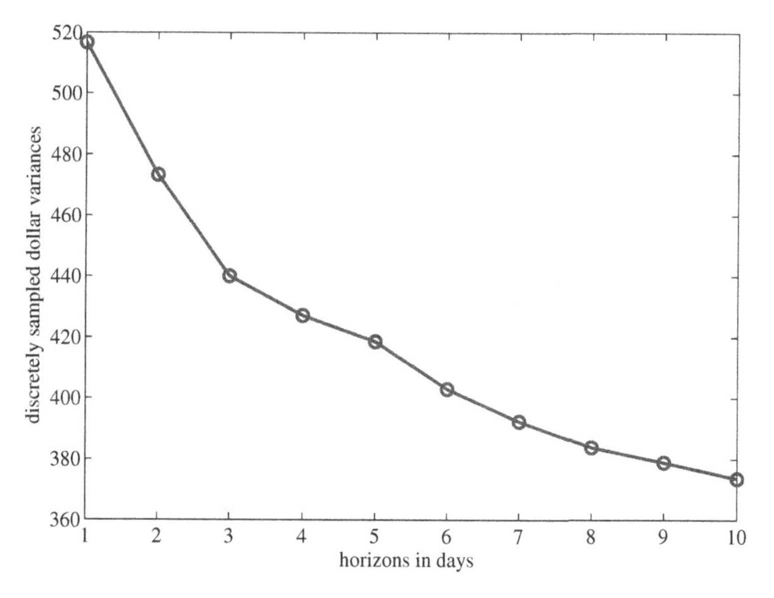

Fig. 3.1: The annualized discrete sampling of the dollar variances of SPY at different horizons for the period Jan 1, 2006 - Sep 5, 2013.

We can easily discretize the previous equation as

$$S_T^2 - S_0^2 \approx \sum_{j=1}^{[T/h]} 2S_{(j-1)h}\left[S_{jh} - S_{(j-1)h}\right] + \left[S_{jh} - S_{(j-1)h}\right]^2 \tag{3.13}$$

and by differencing the two discretized equations at different frequencies (h and H) we get

$$\begin{aligned}&\sum_{j=1}^{[T/h]} \left[S_{jh} - S_{(j-1)h}\right]^2 - \sum_{j=1}^{[T/H]} \left[S_{jH} - S_{(j-1)H}\right]^2 \\ &\approx \sum_{j=1}^{[T/H]} 2S_{(j-1)H}\left[S_{jH} - S_{(j-1)H}\right] - \sum_{j=1}^{[T/h]} 2S_{(j-1)h}\left[S_{jh} - S_{(j-1)h}\right],\end{aligned} \tag{3.14}$$

where the equality holds as long as $[T/h]$ and $[T/H]$ are both integers.

A nice feature about this discretization is that it remains an equality rather than an approximation as in Eq. (3.10), which was simply the Taylor approximation of log function up to the second term. Compared to the earlier case, there are two easily observed differences: (1) the former strategy keeps a constant dollar amount of the underlying stock, whereas the latter keeps a dollar amount (S_t shares) of stocks. (2) the former strategy can be considered asymmetric from a leverage perspective. Suppose S_1 and S_2 are the prices of a stock at time $t = 1$ and $t = 2$ and initially they both hold S_1 shares of the stock. The changes in shares invested for this period for the former strategy is $\frac{(S_1-S_2)S_1}{S_2}$ and those for the latter one is $S_1 - S_2$. If $S_2 = \Delta + S_1$ and $\Delta > 0$, that is, the stock price increases, then the latter strategy will short Δ shares but the former will short slightly less, as $\frac{S_1}{S_2} < 1$. On the other hand, if $\Delta < 0$, i.e. the stock price decreases, then the latter strategy will long $|\Delta|$ shares, whereas the former will long slightly more, as $\frac{S_1}{S_2} > 1$.

Before moving on to the next section, we would like to discuss several observations. (1) The way we formulate these strategies make them contrarian. Without making it too technical, we may see this by assuming $H = 2h$ and $T/h = 2N$, where N is an integer. The profit and loss from Eq. (3.14) can be written as

$$\begin{aligned}&\sum_{j=1}^{[T/h]} \left[S_{jh} - S_{(j-1)h}\right]^2 - \sum_{j=1}^{[T/H]} \left[S_{jH} - S_{(j-1)H}\right]^2 \\ &= -2\sum_{j=1}^{N} \left[S_{2jh} - S_{(2j-1)h}\right]\left[S_{(2j-1)h} - S_{(2j-2)h}\right].\end{aligned} \tag{3.15}$$

This says that, for every 2 h-intervals, if the price first increases then decreases or the price first decreases then increases, we capture the product of the absolute changes. The more frequently the prices change directions, the more profit we gain. (2) This strategy is done periodically. Suppose $H = mh$, then we can carry out this strategy on an m-day sub-period basis. For instance, we can do this by weeks - open trades on Mondays and close trades on Fridays- so that weekend effects are avoided. (3) We have implicitly used days as the frequency unit for this strategy; however, it should work in a high frequency environment, too, as long as the variance ratio statistics are significant and costs are within tolerance. (4) We have discretized the stochastic integrals by unit time intervals, however, this is not necessary. For instance, we can use a stopping time $\tau = \inf_{t>0}\{|S_t - S_0| < \epsilon \text{ and } t < \eta\}$, where ϵ and η are sufficiently small, to decide the next rebalance point.

3.3 The effect of autocorrelation on the strategy performance

It is straightforward to see that when the asset prices follow random walks, the differences between volatilities using different frequency returns should not differ. We will show discrete sampling variances from continuous-time stochastic process such as Merton jump-diffusion model are not capable of fitting empirical discrete variance behaviors, but time series models which specify serial correlation in asset prices are.

3.3.1 *Merton jump-diffusion model*

We first consider the Merton jump-diffusion model. The dynamics of this model is given by

$$\frac{dS_t}{S_t} = (r - \lambda m)dt + \sigma dW_t + dJ_t, \tag{3.16}$$

where $J_t = \sum_{i=1}^{N_t}(Y_j - 1)$ and N_t is a Poisson process with rate λ and Y_j is the relative jump size in the stock price. Y_j is log-normally distributed with parameters a and b^2, that is $LN[a, b^2]$, and m is the mean proportional size of jump, i.e. $E(Y_j - 1) = m$. The parameters a, b and m are related to each other via the following equation

$$e^{a+\frac{b^2}{2}} = m + 1. \tag{3.17}$$

When $\lambda = 0$, this model is reduced to the Black-Scholes model.

Compared to the model without jumps, the addition of the jump factor results in one extra term with $\log Y_i$ in the right-hand side of the next equation

$$\frac{2}{T}\left(\int_0^T \frac{dS_t}{S_t} - \log\frac{S_T}{S_0}\right) = \frac{1}{T}\int_0^T \sigma^2 dt + \frac{1}{T}\left(\sum_{i=1}^{N(T)} \log^2 Y_i\right). \tag{3.18}$$

The expectation of the right-hand side of the equation is the continuous time variance, which equals $\sigma^2 + (a^2 + b^2)\lambda$.

Prop 3.1. If we compute the discretized variance at a time step of Δt and define its value as $F(\Delta t)$, then it can be shown [Broadie and Jain (2008)] that the discretized variance is equal to the continuous-time variance plus multiple of Δt, that is,

$$\begin{aligned} F(\Delta t) &= \sigma^2 + (a^2 + b^2)\lambda \\ &+ \frac{(r - \lambda m - \frac{1}{2}\sigma^2)^2 T + a^2\lambda^2 T + 2(r - \lambda m - \frac{1}{2}\sigma^2)a\lambda T}{n}, \end{aligned} \tag{3.19}$$

where $n = T/\Delta t$. Further, taking the derivative of $F(\Delta t)$, we easily obtain

$$F'(\Delta t) = (r - m\lambda + a\lambda - \frac{1}{2}\sigma^2)^2. \tag{3.20}$$

The proof of this proposition can be found in the appendix of this chapter.

Since the derivative of $F(\Delta t)$ in Δt is always non-negative, it implies that the effect of discretization and jumps in prices generated from Merton jump-diffusion model always gives the realized variances greater than their continuous-time limit, with a discrepancy that is linear in Δt; downward sloping variance ratios against the horizon of returns are not possible under these two factors. Therefore, a random walk based models with jumps is not consistent with our empirical observations on variance ratios for the period 2006 - 2013 seen in Chapter 1. We next resort to autoregressive time series models.

3.3.2 *A Stationary Model*

Suppose we have an AR(1) process denoted by X_n satisfying

$$X_n = aX_{n-1} + b + w_n, \tag{3.21}$$

where $a \in (0, 1), b$ are constants, w_n are IID random variables with mean 0 and standard deviation σ. It is elementary to show

$$\mu = E(X_n) = \frac{b}{1-a}$$

$$\gamma_k = Cov(X_n, X_{n-k}) = \frac{\sigma^2 a^k}{1-a^2},$$

for all n and $k = 0, 1, 2, ...$, using its MA(∞) representation

$$X_n = \frac{b}{1-a} + w_n + aw_{n-1} + a^2 w_{n-2} + a^3 w_{n-3} + \dots . \tag{3.22}$$

Denote the one step difference of X_n by Y_n, that is,

$$Y_n = X_n - X_{n-1}. \tag{3.23}$$

Then,

$$Y_n = w_n + (a-1)(w_{n-1} + aw_{n-2} + a^2 w_{n-3} + ...). \tag{3.24}$$

Since we will be dealing with quantities with Y_n's mostly next, without loss of generality, we assume $b = 0$ to simplify the computation. In this case, we have the system as

$$X_n = aX_{n-1} + w_n, \tag{3.25}$$

where $a \in (0, 1)$ is constant, w_n are IID random variables with mean 0 and standard deviation σ. Moreover,

$$X_n = w_n + aw_{n-1} + a^2 w_{n-2} + a^3 w_{n-3} + \dots . \tag{3.26}$$

Thus, for $k \geq 0$

$$\mu = E(X_n) = 0$$

$$\gamma_k = E[X_n X_{n-k}] = \frac{\sigma^2 a^k}{1-a^2}.$$

If X_n represents price dynamics, then Y_n will denote price differences. If X_n is the log price, then Y_n represent the log returns. As a result, the behaviors of Y_n are of great interest. We derive several properties of Y_n and summarize them in the following proposition.

Prop 3.2.

I The mean of Y_n is zero,

II Autocovariance defined by Γ_k is

$$\Gamma_k = E\left[Y_n Y_{n-k}\right] = -\frac{\sigma^2 a^{k-1}(1-a)}{1+a} < 0, \tag{3.27}$$

where $k \geq 1$.

III Variance defined by Γ_0 is

$$\Gamma_0 = Var Y_n = \frac{2\sigma^2}{1+a}. \tag{3.28}$$

IV Variance of $Y_1 + ... + Y_n$ is

$$Var[Y_1 + ... + Y_n] = \frac{2\sigma^2(1-a^n)}{1-a^2}. \tag{3.29}$$

The proof of this proposition is in the appendix of this chapter.

Define

$$Q(n,a) = \frac{Var[Y_1 + ... + Y_n]}{n} = \frac{2\sigma^2(1-a^n)}{n(1-a^2)},$$

and regard it as normalized variances. We investigate its monotonicity by Figure 3.2, which plots $Q(n,a)$ for various n levels ($n = 1, 2, 4, 16$) across all $a \in (0,1)$. We see that as n increases, $Q(n,a)$ shifts lower for all a in $(0,1)$. However, the monotonicity of $Q(n,a)$ in a changes from monotone decreasing to a convex function which first declines and then rises as a increases from 0 to 1.

Now, define the P&L of our systematic strategy as Z_k, which is the difference between the variance sampled using k-period price differences and that using 1-period price differences,

$$Z_k = (X_k - X_{k-1})^2 + (X_{k-1} - X_{k-2})^2 + ... + (X_1 - X_0)^2 - (X_k - X_0)^2,$$

then we can show

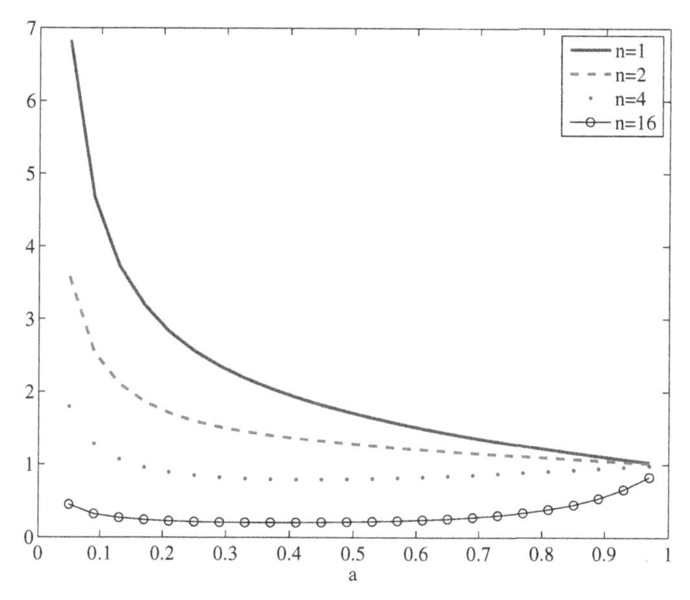

Fig. 3.2: The normalized variance $Q(n, a)$ for various n levels across $a \in (0, 1)$. The y-axis represents the level of $Q(n, a)$.

Prop 3.3. If we denote

$$f(\sigma, k, a) = EZ_k \tag{3.30}$$

$$s(\sigma, k, a, \eta)^2 = EZ_k^2 - (EZ_k)^2, \tag{3.31}$$

as the mean and variance of the P&L, where η is the kurtosis of w_t, then

$$f(\sigma, k, a) = \sigma^2 \left(\frac{2k}{1+a} - 2\frac{1-a^k}{(1-a)(1+a)} \right). \tag{3.32}$$

In addition, we can show the following properties:

I

$$\lim_{a \to 1^-} f(\sigma, k, a) = 0, \tag{3.33}$$

which is consistent with our understanding of the case where X_n follows a random walk.

II $f(\sigma, k, a)$ is increasing in k for all $k = 1, 2, 3, ...$

III $f(\sigma, k, a)$ decreases in a for $k \geq 2$.

IV Since $f(\sigma, k, a)$ is always positive, it increases as σ increases.

The analytical form of $s(\sigma, k, a, \eta)^2$ is rather complex and it can be found in the appendix together with properties of the function's sensitivities in its parameters illustrated graphically. The proof of this proposition is also in the appendix of this chapter.

The function $f(\sigma, k, a)$ is in fact a constant multiple of the P&L of the earlier strategy if the corresponding underlying asset price follows an AR(1) process. Recall Eq. (3.14).

$$\sum_{j=1}^{[T/h]} \left[S_{jh} - S_{(j-1)h}\right]^2 - \sum_{j=1}^{[T/H]} \left[S_{jH} - S_{(j-1)H}\right]^2$$
$$\approx \sum_{j=1}^{[T/H]} 2S_{(j-1)H} \left[S_{jH} - S_{(j-1)H}\right] - \sum_{j=1}^{[T/h]} 2S_{(j-1)h} \left[S_{jh} - S_{(j-1)h}\right].$$

The left-hand side shows the total P&L of this trade for the period $[0, T]$ and the right-hand side shows us the tactic we should adopt. Suppose $H = kh$ and $T/h = kN$. Denote $X_j = S_{jh}$ and $V_k = \sum_{j=1}^{[T/H]} \left[S_{jH} - S_{(j-1)H}\right]^2$. Then, we may write

$$V_k = \sum_{j=1}^{N} \left[X_{jk} - X_{(j-1)k}\right]^2 \tag{3.34}$$

and the left-hand side of Eq. (3.14) as

$$\begin{aligned} V_1 - V_k &= \sum_{j=1}^{kN} \left[X_j - X_{(j-1)}\right]^2 - \sum_{j=1}^{N} \left[X_{jk} - X_{(j-1)k}\right]^2 \\ &= \sum_{j=1}^{N} \left[X_{jk} - X_{jk-1}\right]^2 + \left[X_{jk-1} - X_{jk-2}\right]^2 + \\ &\quad \ldots + \left[X_{jk-k+1} - X_{jk-k}\right]^2 - \left[X_{jk} - X_{(j-1)k}\right]^2. \end{aligned} \tag{3.35}$$

Thus,

$$\begin{aligned} E(V_1 - V_k) &= E\sum_{j=1}^{N} \left[X_{jk} - X_{jk-1}\right]^2 + \left[X_{jk-1} - X_{jk-2}\right]^2 + \\ &\quad \ldots + \left[X_{jk-k+1} - X_{jk-k}\right]^2 - \left[X_{jk} - X_{(j-1)k}\right]^2 \\ &= -N\ EZ_k \end{aligned}$$

$$= Nf(\sigma, k, a). \tag{3.36}$$

Hence, our systematic strategy can be decomposed into many k-interval units. Since $f(\sigma, k, a) > 0$, the higher frequency variance must be greater than lower frequency variance under this simplistic stationary model. The parameter a measures the strength of the mean reversion of this process. The closer a is to 0, the faster the process mean-reverts. Since $f(\sigma, k, a)$ decreases with a, the strategy is more profitable when the price series mean-reverts more frequently. Moreover, the P&L for this k-unit period is larger if we have more volatility in the innovations.

To compare with the Merton's jump-diffusion model, we plot the discrete sampling of variances obtained by Monte Carlo simulation of the AR(1) process for various levels of the parameter a. Figure 3.3 compares

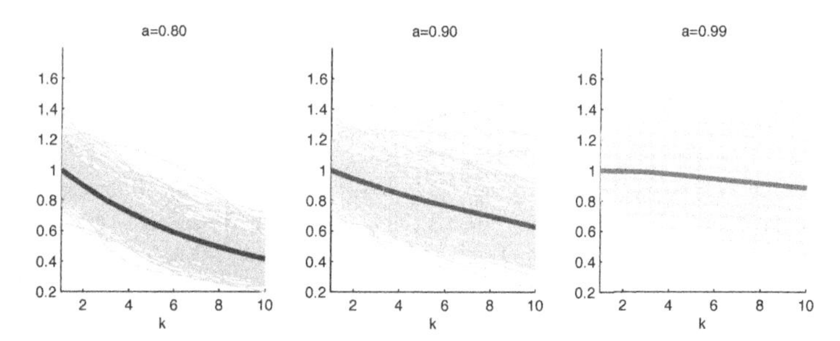

Fig. 3.3: The discretely sampled variances of Y_n at different horizons (k) for different values of a are compared The variation approximations are normalized for comparison so that the ones associated $k = 1$ is 1. The thin lines are calculated from each sample path of the corresponding AR(1) process. The thick line shows their average in each case.

the discretely sampled variances of Y_n at different horizons (k) for different values of a. The variation approximations are normalized for comparison so that the ones associated with $k = 1$ is 1. In the simulations of this figure, we have assumed that the innovations w_n follows a normal distribution with $\sigma = 0.15$. The thin lines are calculated from each sample path of the corresponding AR(1) process. The thick line shows their average in each case. The shapes of the lines are consistent with our results on $f(\sigma, k, a)$ as well as the empirical observations for SPY in the period 2006-2013.

Figure 3.4 reports the annualized risk-adjusted P&L, i.e. $\sqrt{\frac{252}{k}} \frac{f(\sigma,k,a)}{s(\sigma,k,a,\eta)}$

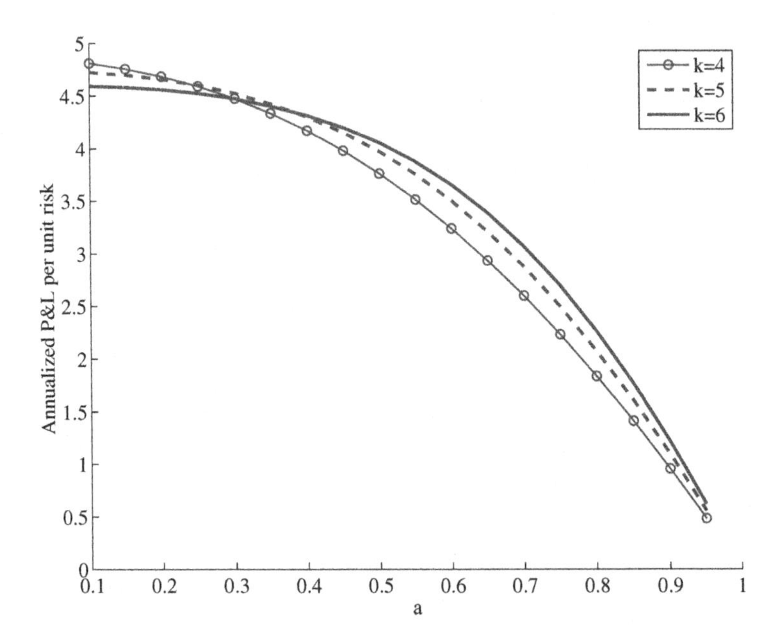

Fig. 3.4: The risk adjusted P&L $\sqrt{\frac{252}{k}}\frac{f(0.3,k,a)}{s(0.3,k,a,4)}$ for $k = 4, 5$ and 6 across $a \in (0, 1)$, where we have indicated $\eta = 4$ and $\sigma = 0.3$. The function is decreasing in a and concave downward. As k increases, the function curve shifts upward and its curvature increases.

for $k = 4, 5$ and 6 across $a \in (0, 1)$, where we have indicated $\eta = 4$ and $\sigma = 0.3$. The function is decreasing in a and concave downward, which means that the gain per unit risk of this strategy is larger for an AR(1) process that exhibits stronger mean-reversion. As $a \to 1^-$, the P&L converges to 0, corresponding to the case where the underlying asset price follows a random walk. The figure also shows that as k increases, the function curve shifts upward and its curvature increases. This suggests longer sub-period offers better profitability, provided a is constant for various k's. The stronger curvature suggests that per unit decrease in a will improve the risk-adjusted gains faster if k is bigger. In addition, near the $a = 0$ end, we see that larger k reduces the risk-adjusted gain. This is more likely due to the fact that volatility of the P&L increases a lot more than the jump in the gain itself as k rises, however, the effect here is less significant compared to the $a = 1$ end and in practice, most asset prices have parameter a close to 1. In addition, we have found that the volatility level, σ, in the innovations does not affect the risk-adjusted P&L.

We also test our strategy on simulated AR(1) processes for the case of $k = 5$ with an initial cash investment of 1000. We simulate 500 independent and identically distributed Gaussian random variables with mean 0 and standard deviation $\sigma = 0.15$ and use this set of noise to create AR(1) time series of length 500 starting at the level of 100 for $a = 0.7, 0.8, 0.9$ and 1.0. We consider these four time series as price series, apply our strategy on these paths and collect their annualized Sharpe ratios respectively. For the moment, we ignore the practical details such as cost, leverage, etc., as our goal is to understand the effects of parameter a in the specified auto regressive model. Figure 3.5 reports the cumulative performances for the four cases whose prices are generated from one particular simulation of Gaussian noises. It is evident from the cumulative performances that the case where $a = 0.7$ offers the best result. The Sharpe ratios for the price series with $a = 0.7, 0.8, 0.9$ and 1.0 are 3.41, 2.31, 0.85 and -0.78 respectively.

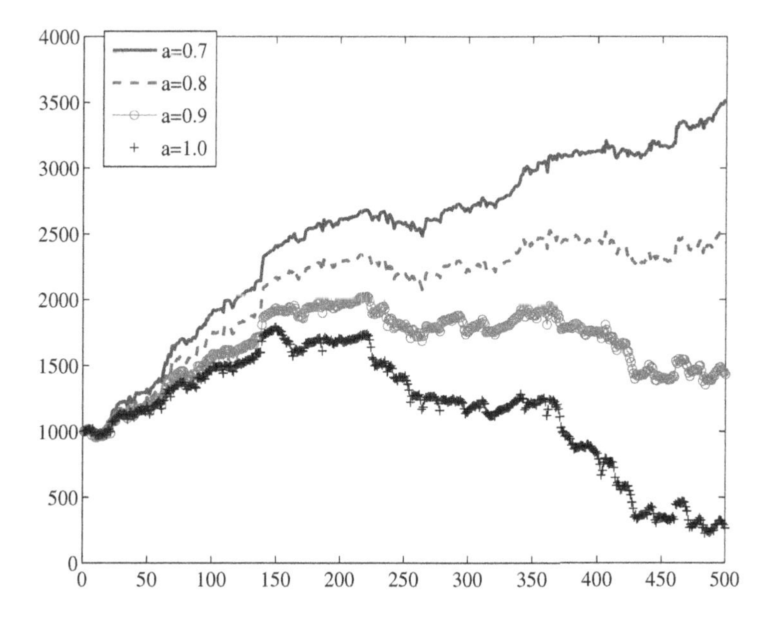

Fig. 3.5: The cumulative performances of the arbitrage strategy on four AR(1) price series generated with $a = 0.7, 0.8, 0.9$ and 1.0 with Gaussian noise distributed as $N(0, 0.15^2)$. The Sharpe ratios for the price series with $a = 0.7, 0.8, 0.9$ and 1.0 are 3.41, 2.31, 0.85 and -0.78 respectively.

Of course, one particular simulation is not sufficient to prove the aggregate property we propose. We also investigate the distributional behaviors of these strategies. To do so, we repeated this simulation for 100 times. Figure 3.6 reports the Sharpe ratios in the form of a 100% stack chart for all 100 cases, where for each simulation the four Sharpe ratios corresponding to the four levels of a are normalized so that their sum is 1. The x-axis represents the number of simulations. We see that the Sharpe ratios for $a = 0.7$ across the 100 simulations have overall the largest weights and $a = 1.0$ the smallest. In quite a number of scenarios, we see that it has negative Sharpe ratios, which never occur to $a = 0.7$ and 0.8 in the 100 simulated situations. The averaged Sharpe ratios for $a = 0.7, 0.8, 0.9$ and 1.0 are 3.00, 2.17, 1.24 and 0.36 respectively. This experiment shows that not only the P&L is larger for AR(1) process with smaller a, the corresponding risk adjusted return has the same property.

3.3.3 *A Model with Trend*

The previous section provides analysis for a stationary series of prices. However, in practice, stock prices typically experience trends. As a result, we consider the model of the form

$$\begin{aligned} S_n &= d + cn + X_n \\ X_n &= aX_{n-1} + b + w_n. \end{aligned} \tag{3.37}$$

Define

$$Z_k = (S_k - S_{k-1})^2 + (S_{k-1} - S_{k-2})^2 + (S_1 - S_0)^2 - \dots - (S_k - S_0)^2 \tag{3.38}$$

and

$$g(\sigma, k, a, c) = EZ_k, \tag{3.39}$$

then in the next proposition we derive the close-form representation of g.

Prop 3.4. Under the model with trend,

$$\begin{aligned} g(\sigma, k, a, c) &= f(\sigma, k, a) - k(k-1)c^2 \\ &= \sigma^2 \left(\frac{2k}{1+a} - 2\frac{1-a^k}{(1-a)(1+a)} \right) - k(k-1)c^2. \end{aligned} \tag{3.40}$$

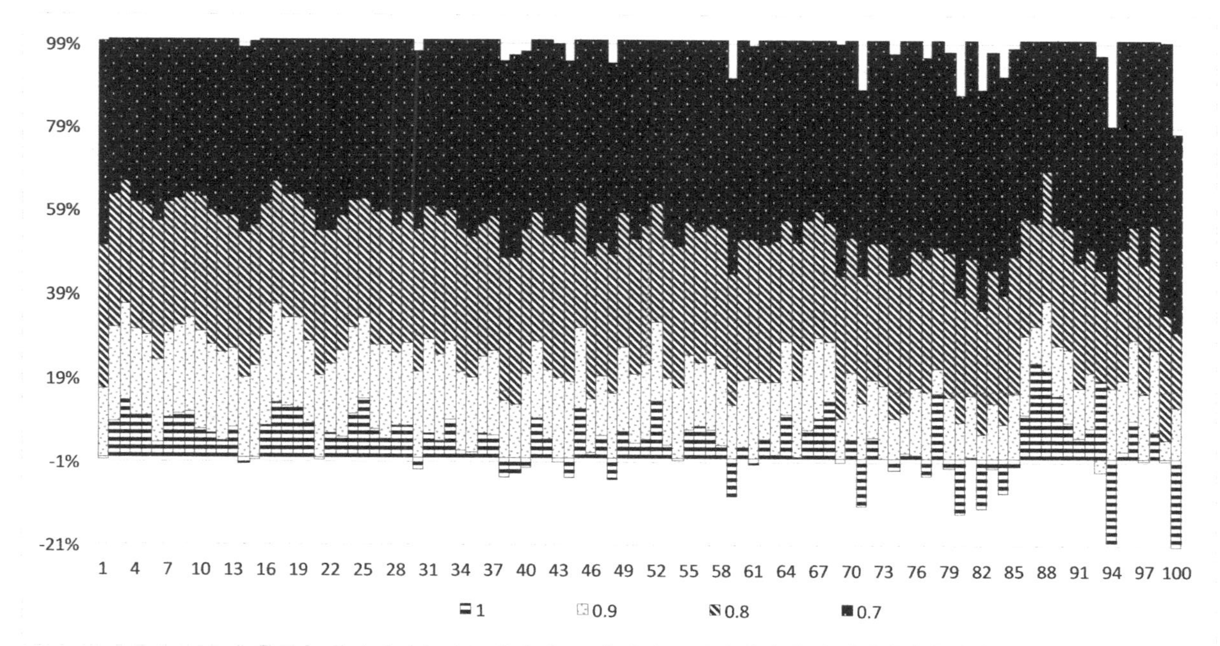

Fig. 3.6: We simulate 500 independent and identically distributed Gaussian random variables with mean 0 and standard deviation $\sigma = 0.15$ and use this set of noise to create AR(1) time series of length 500 starting at the level of 100 for $a = 0.7, 0.8, 0.9$ and 1.0. We consider these four time series as price series, apply our strategy on these paths and collect their annualized Sharpe ratios respectively. Repeat this for 100 times. This figure reports the Sharpe ratios in the form of a 100% stack chart for all 100 cases, where for each simulation the four Sharpe ratios corresponding to the four levels of a are normalized so that their sum is 1. The averaged Sharpe ratios for $a = 0.7, 0.8, 0.9$ and 1.0 are 3.00, 2.17, 1.24 and 0.36 respectively.

The proof of this proposition is in the appendix of this chapter.

To understand the monotonicity of $g(\sigma, k, a, c)$ in k, we compute

$$\begin{aligned} & g(\sigma, k+1, a, c) - g(\sigma, k, a, c) \\ & = -(k+1)kc^2 + k(k-1)c^2 + f(\sigma, k+1, a) - f(\sigma, k, a) \\ & = -2kc^2 + 2\sigma^2 \frac{1-a^k}{1+a}. \end{aligned} \tag{3.41}$$

Hence, $g(\sigma, k, a, c)$ increases in k if and only if $\sigma^2(1-a^k) - (1+a)kc^2 > 0$. However, it is easy to see for k very large, the whole expression will remain negative, therefore, in such cases, $g(\sigma, k, a, c)$ will decrease with k. In addition, following the properties of $f(\sigma, k, a)$ and trivial arguments, it is easy to show that $g(\sigma, k, a, c)$ decreases in a, c and increases in σ. Whether $g(\sigma, k, a, c)$ is positive or negative is more involved and will depend on the independent variables. Therefore, trending in the time series can hurt the performance of our strategy. However, it is worth pointing out that empirically the trending coefficient c can be quite small.

3.4 Backtesting

In this section, we backtest our strategy on ETFs, using Eq. (3.13) with an initial cash budget of \$100 and assuming $H = 5h$, that is one day versus five days. We also operate on a five-day cycle, that is, we enter the trades on the first day and rebalance the positions according to Eq. (3.13) daily, but on the fifth day we close all positions and realize the gain and loss. On the sixth day, we restart the strategy and continue until the last day of backtesting period is reached. The trades are assumed to be executed before market close so that the adjusted close prices are used. We will also consider a commission cost of 5 basis points per share and a borrowing cost of 100 basis points annually.

3.4.1 *A SPY variance strategy for 2006-2013*

We consider the ETF with the ticker SPY for illustration. SPY is the SPDR S&P500 Index ETF. One of the earliest and most liquid exchange traded funds. We will demonstrate how the systematic strategy performs when SPY is used as the underlying asset. Strategy performance related statistics are plotted in Figures 3.7, 3.8, 3.9 and 3.10 for each individual year of the period 2007-2009 and 2011.

Each of these figures contain four panels. The top panel contains two curves, the dashed line always underperforms the solid one since it accounts for trading costs whereas the other does not. The second panel shows the performance of the underlier for the same period with initial level normalized to 100. The third panel shows the monthly moving average of leverage computed on a daily basis as the absolute equity exposure over the total capital available of the trader. For instance, at the beginning the total capital the trader owns is 100 dollars. The bottom panel shows the drawdown and maximum drawdown presented as the solid and dashed lines respectively. Maximum drawdown is always greater than or equal to the drawdown levels at each point of time. We test the strategy in a way that the leverage levels are typically around 1.

In addition, Table 3.1 reports the key performance statistics including annualized returns, volatilities and Sharpe ratios of the strategy for SPY for each year of the period 2006-2013. The sample period contains both the pre-crisis, criss and recovery economic periods, which helps us better understand how this proposed strategy behaves in time. For simplicity, we will not separate the data history into in-sample and out-of-sample segments, instead we simply consider the full sample scenario. The goal is to understand how the performance of the strategy conforms with our earlier analysis. The methodology here is probably in its simplest form, though many sophisticated extensions can be rather easily achieved to mitigate risk or extract additional alpha.

Fig. 3.7: The quantities associated with strategy performance for the year 2007. The top panel contains two curves, the dashed line always underperforms the solid one since it accounts for trading costs whereas the other does not. The second panel shows the performance of the underlier for the same period with initial level normalized to 100. The third panel shows the monthly moving average of leverage computed on a daily basis as the absolute equity exposure over the total capital available of the trader. For instance, at the beginning the total capital the trader owns is 100 dollars. The bottom panel shows the drawdown and maximum drawdown presented as the solid and dashed lines respectively. Maximum drawdown is always greater than or equal to the drawdown levels at each point of time.

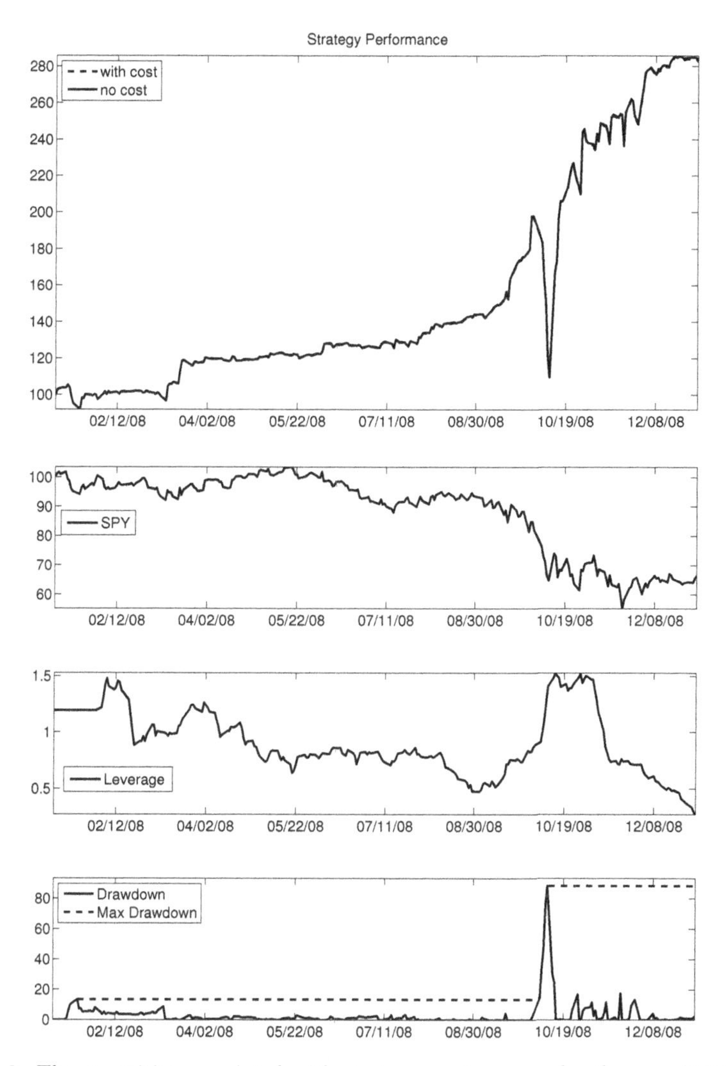

Fig. 3.8: The quantities associated with strategy performance for the year 2008. The top panel contains two curves, the dashed line always underperforms the solid one since it accounts for trading costs whereas the other does not. The second panel shows the performance of the underlier for the same period with initial level normalized to 100. The third panel shows the monthly moving average of leverage computed on a daily basis as the absolute equity exposure over the total capital available of the trader. For instance, at the beginning the total capital the trader owns is 100 dollars. The bottom panel shows the drawdown and maximum drawdown presented as the solid and dashed lines respectively. Maximum drawdown is always greater than or equal to the drawdown levels at each point of time.

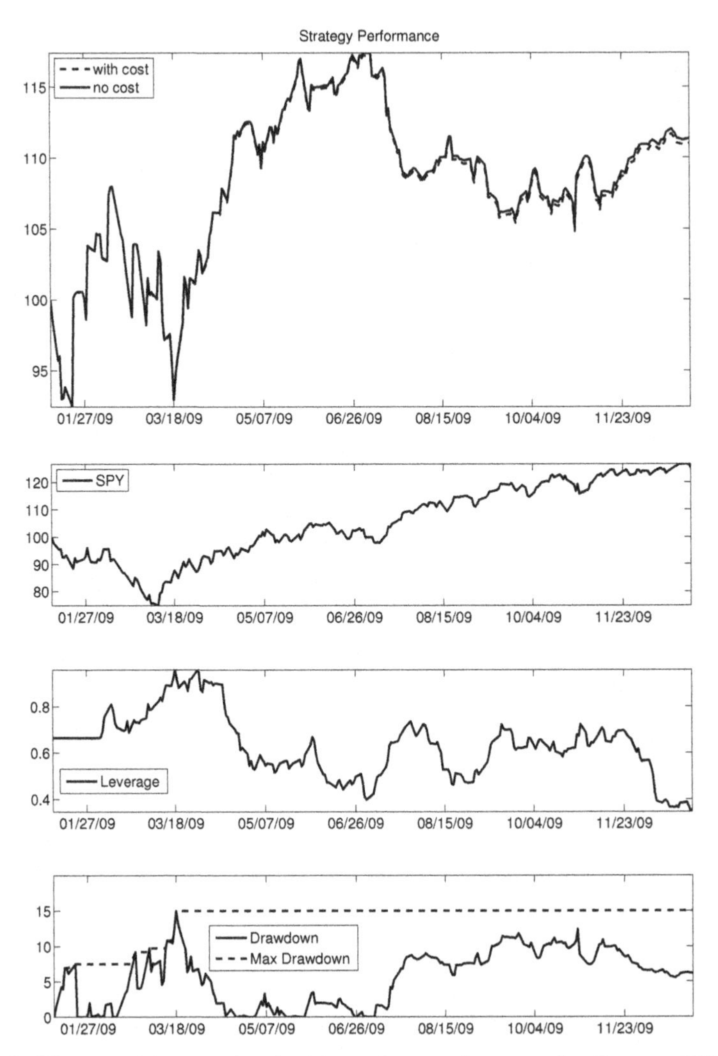

Fig. 3.9: The quantities associated with strategy performance for the year 2009. The top panel contains two curves, the dashed line always underperforms the solid one since it accounts for trading costs whereas the other does not. The second panel shows the performance of the underlier for the same period with initial level normalized to 100. The third panel shows the monthly moving average of leverage computed on a daily basis as the absolute equity exposure over the total capital available of the trader. For instance, at the beginning the total capital the trader owns is 100 dollars. The bottom panel shows the drawdown and maximum drawdown presented as the solid and dashed lines respectively. Maximum drawdown is always greater than or equal to the drawdown levels at each point of time.

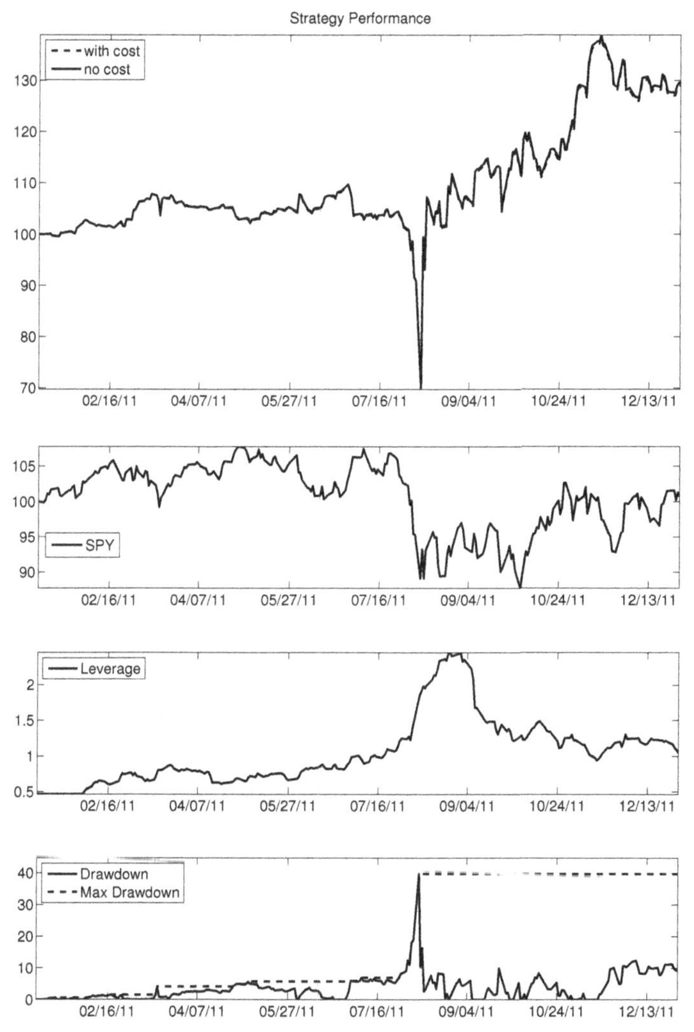

Fig. 3.10: The quantities associated with strategy performance for the year 2011. The top panel contains two curves, the dashed line always underperforms the solid one since it accounts for trading costs whereas the other does not. The second panel shows the performance of the underlier for the same period with initial level normalized to 100. The third panel shows the monthly moving average of leverage computed on a daily basis as the absolute equity exposure over the total capital available of the trader. For instance, at the beginning the total capital the trader owns is 100 dollars. The bottom panel shows the drawdown and maximum drawdown presented as the solid and dashed lines respectively. Maximum drawdown is always greater than or equal to the drawdown levels at each point of time.

No cost				With cost			
Year	Return	Volatility	Sharpe ratio	Year	Return	Volatility	Sharpe ratio
2006	0.04	0.05	0.84	2006	0.04	0.05	0.8
2007	0.33	0.2	1.64	2007	0.33	0.2	1.62
2008	1.28	0.71	1.81	2008	1.28	0.71	1.8
2009	0.13	0.2	0.64	2009	0.13	0.2	0.63
2010	0.1	0.21	0.48	2010	0.1	0.21	0.46
2011	0.42	0.58	0.72	2011	0.41	0.58	0.71
2012	0.06	0.13	0.43	2012	0.05	0.13	0.39
2013	0.09	0.13	0.64	2013	0.08	0.13	0.6

Table 3.1: The annualized returns, volatilities and Sharpe ratios of the strategy for SPY for each year of the period 2006-2013.

From both the table and the figures, we see that the strategy performed very well in the years of 2007 and 2008 during the financial crisis, when the underlying ETF performed rather poorly. For instance, in 2007, the strategy has an annualized return of 33%, annualized volatility of 20% and a Sharpe ratio of 1.64 with a leverage level around 1, whereas the return on SPY was negative. This is very likely due to some of the factors we investigated in the earlier chapters. (1) During the crisis, the return volatility is elevated. This, as seen in Chapter 2, is often associated with lower levels of return autocorrelations. (2) The interest rates are close-to-zero; the autocorrelation fell below zero and became statically significant. As a result, the mean-reverting behavior of the security offered significant gains to the strategy. However, as the economy started to recover, the strength of the strategy declined. As in the year 2010, the volatility of SPY has considerably reduced and the variance strategy only gained a Sharpe ratio of 0.46. Nonetheless, shocks from the international markets such as the burst of the European debt crisis in 2010 and the US downgrade in 2011 added uncertainty to the US economy, offering less significant but viable profits to our strategy. In 2012 and 2013, the ETF itself achieved tremendous growth, leaving lackluster performances to the systematic strategy.

From the figures, we do see, however, that the strategy can occasionally suffer from very large drawdowns[2]. During high volatility market environments, such as crisis, there tend to be significant price displacements, which cause the strategy to buy and sell more shares each day; whereas recovery or boom periods, prices do not fluctuate much, the number of shares daily traded is less. The strategy performs well when the prices oscillate more

[2]Drawdowns are defined in the Appendix.

frequently. If the prices are trending in one direction for a relatively long period, a rather large drawdown is very likely to occur. In the case where the trending period is longer than the sub-strategy cycle, the strategy has to unwind its positions, realizing the trades in that period as a loss. On the other hand, if the asset price reverts in direction before the termination of the cycle, the drawdown will be recovered, either partially or entirely, and the P&L is not significantly affected, yet the risk associated will be magnified. As a result, a positive but low risk-adjusted return should be expected. For instance, if we look at the period around the US downgrade event in August 2011, the cumulative performance of the strategy experiences a large downward spike reaching almost 45% resulted from consecutive drops in SPY's price. A relatively comforting side of the story is that the contrarian bets from the strategy soon pushed the curve back to approximately the same level before the spike. That is, the drawdown duration is relatively short. Even though the total annualized gain is over 41% of the year 2011, the high volatility of 58% associated resulted in a risk-adjusted return of only 0.71.

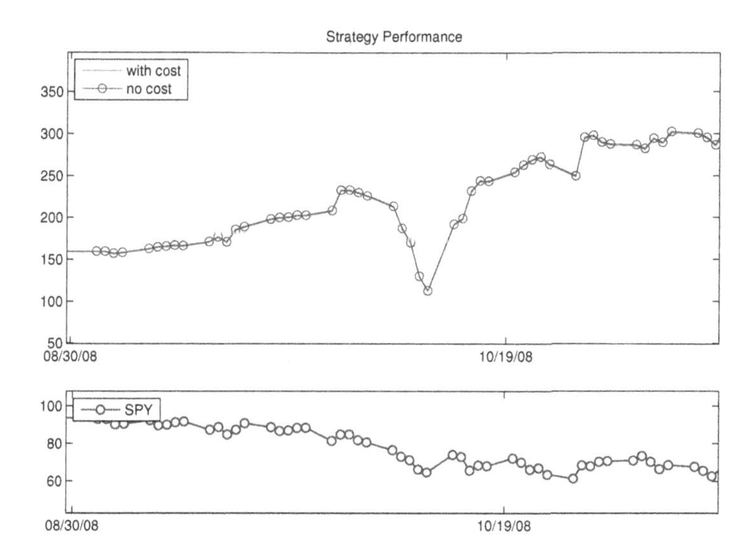

Fig. 3.11: The strategy performance on SPY for the week October 6-10, 2008 in the upper panel, whereas the levels of SPY is plotted in the lower panel. For the purpose of comparison, the levels of the strategy and SPY are both normalized to 100 on the first trading day of 2008.

Another instance of big drawdowns is the week of October 6-10, 2008. Figure 3.11 plots the zoom-in charts of Figure 3.8 for this particular week. According to the news, this period is the worst week for the stock market in 75 years. The Standard & Poor's 500 index loses 18.2%, its worst week since 1933, down 42.5% since its own high on October 9, 2007. The following are some of the important events took place for that week.

- October 6: The Federal Reserves announced that it would provide as much as $900 billion in cash loans to banks and will start to pay interest on commercial banks' reserves.
- October 7: The central bank invoked emergency powers to lend money to companies outside the financial sector and buy up commercial paper. The Fed signals to cut interest rates.
- October 8: The world's major central banks lowered their benchmark interest rates to halt a collapse of asset prices and a freeze in the credit markets.
- October 9: The Treasury said it was looking to buy stakes in some banks as part of the $700 billion bank bailout law enacted the week before.
- October 10: Stocks, oil, gas and gold prices fell. Credit market remained tight. President Bush said he would host a meeting with G-7 Finance Ministers early Saturday morning.

As the price of SPY continued to drop, the strategy systematically accumulate more and more position in the ETF so that as soon as the price renounces, the strategy would pick up the gain. Indeed, this is exactly what happened the week after. The strategy then recovered from the drawdown and continued to profit for the rest of 2008.

In all cases, the naive transaction costs do not seem to reduce the gain of the strategy too much. Partially, this is because the ETF under consideration is relatively liquid and that a low level of leverage is enough to generate meaningful size of gains. However, if the ETF is hard-to-borrow or high leverage is needed to generate sufficient returns, transaction costs can significantly reduce the performance of the strategy. In addition to commission and borrowing cost, we also attempt to include additional costs, say x basis points per share, that may account for the difference between the close price which we use for signal and the actual realized price when the trades are successfully executed. Table 3.2 shows how the annualized returns, annualized volatility and Sharpe ratios change as the cost x changes. It appears that the additional costs reduce returns but induce more volatility in the performance of the strategy. However, despite these costs, the risk adjusted level of returns remains reasonably good.

x	Return	Volatility	Sharpe ratio
100	1.28	0.72	1.80
300	1.28	0.72	1.79
700	1.27	0.73	1.76

Table 3.2: The annualized returns, volatilities and Sharpe ratios of the strategy for SPY in 2008 for various levels of additional costs, x, in basis points.

When considering faster strategies, the cost aspect is rather important. Many historical market anomalies proposed and backtested on paper were not able to be put into practical trading because of transaction costs. There has also been quite a number of research on relevant topics such as hard-to-borrow securities and market impacts. For instance, Almgren and Chriss (2001) considered the execution of portfolio transactions with the aim of minimizing a combination of volatility risk and transaction costs arising from permanent and temporary market impact. Their simple linear cost model implies that when market impacts are taken into consideration, the liquidation of a portfolio by risk-averse traders should first go fast, however, as the trade sizes increase, the liquidating speed should slow down. Garleanu and Pedersen (2013) derived a closed-form optimal dynamic portfolio policy when trading cost is accounted and security returns are predictable by various mean-reverting signals. They found that the optimal strategy always trades partially toward the current aim, though it is not the same as the efficient market frontier. Further, they implemented the optimal strategy for commodity futures and found exceptional net returns compared to simple benchmarks.

3.4.2 *The SPY variance strategy in 2008 at different horizons*

A separate parameter we yet need to investigate is the sample variance horizon or lag. Recall that we have so far assumed that $H = 5h$. If we vary the ratio between H and h so that it is no longer 5, what can happen? In order to answer this question. We continue to use SPY as the underlying security. Again, we consider the initial capital is \$100. To make things easier, we set all costs to be zero.

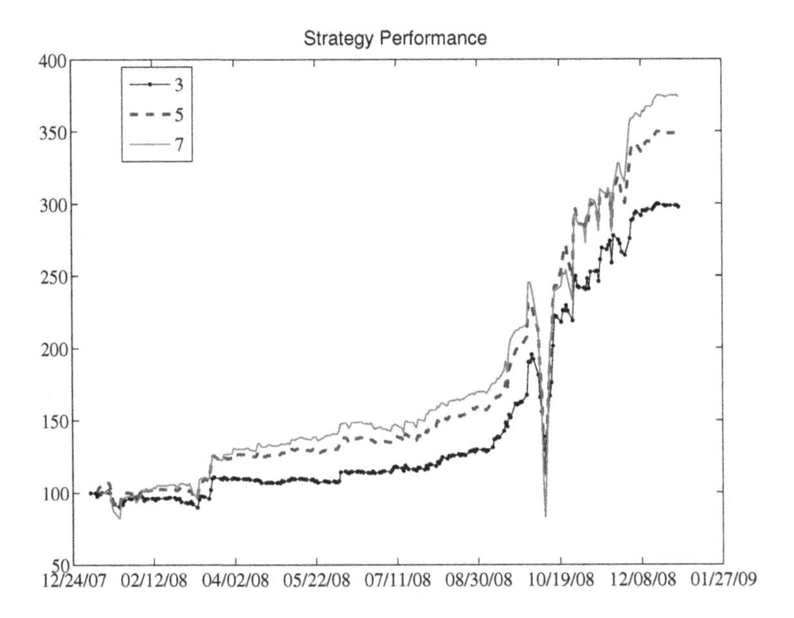

Fig. 3.12: The strategy performance on SPY in 2008 with horizons equal to 3, 5 and 7.

We perform backtesting in the same manner as the previous section. Figure 3.12 shows the cumulative performance of the strategy for three different horizons (3, 5 and 7), so that the trading in practice will be subject to 3-day, 5-day and 7-day cycles respectively. Although the shape of these strategy cumulative returns look rather similar, it appears that the bigger the horizon, the higher the performance, as the curve denoted by the solid line in the figure obtains the highest cumulative return by the end of 2008. The dashed line for horizon 5 ranks the second, with case horizon equal to 3, the lowest. However, it is worth noting that associated with high returns is high risks. Table 3.3 reports the annualized returns, volatilities and Sharpe ratios of the strategy for SPY in 2008 for the three different horizons. We see that the annualized risk is the highest at horizon 7 at 1.65 and smallest at horizon 3 slightly above 0.5. As a result, when these strategies are measured by the risk-adjusted return, the strategy at horizon 3 is actually the best one. It has a Sharpe ratio of 2.41, whereas the strategy at horizon 7 only has a Sharpe ratio equal to 1.35 for this particular year regardless of the cost.

horizon	annual return	annual risk	Sharpe ratio
3	1.22	0.52	2.41
5	1.59	0.9	1.8
7	2.19	1.65	1.35

Table 3.3: The annualized returns, volatilities and Sharpe ratios of the strategy for SPY in 2008 for different horizons. The strategy at horizon 3 has the highest risk-adjusted return.

As we have seen from the earlier analysis, another important aspect of the performance of these strategies is their drawdowns. We have also seen that this particular strategy in 2008 suffered significantly from deep drawdowns, even though its Sharpe ratio remains rather good in comparison to the performance of the equity market as a whole. Figure 3.13 plots separately the drawdowns of the strategies at horizons 3, 5 and 7, where the upper panel is for horizon 3, middle for horizon 5 and bottom for horizon 7. The y-axis for all subplots are rescaled to range from 0 to 200. The step functions in dashed lines represent the maximum drawdowns whereas the solid lines represent the regular drawdowns. We see that the horizon-3 strategy has overall the lowest level of drawdowns whereas the horizon-7 strategy has the largest drawdowns. As we have illustrated before, the maximum drawdown occurred for the week of October 6-9, 2008 after the collapse of Lehman Brothers. Hence, for this particular strategy, we have found implementing it at different horizons can offer different risk and return profiles. Apart from this particular week, the drawdowns of these strategies are much more reasonable.

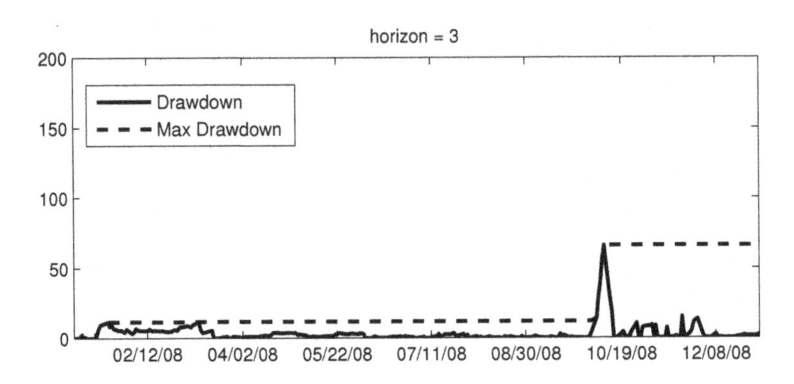

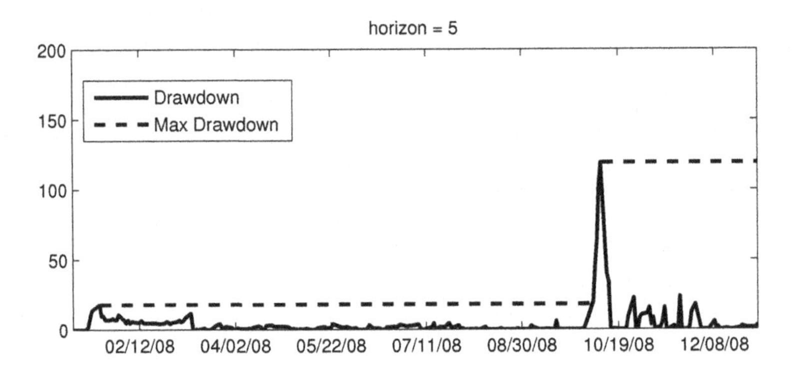

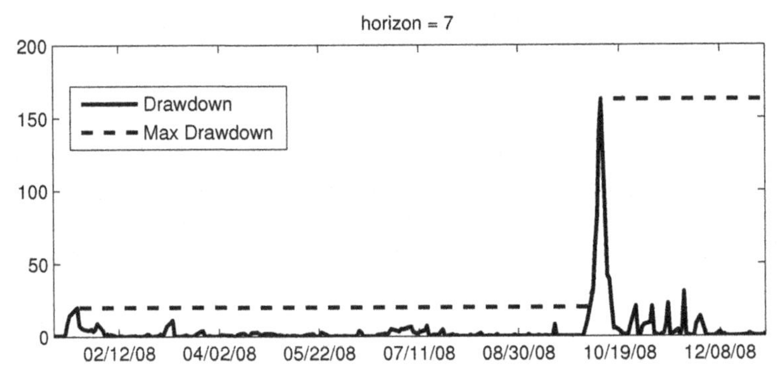

Fig. 3.13: The drawdowns of the strategy performance on SPY in 2008 with horizons equal to 3, 5 and 7, where the upper panel is for horizon 3, middle for horizon 5 and bottom for horizon 7. The y-axis for all subplots are rescaled to range from 0 to 200. The step functions in dashed lines show the maximum drawdowns whereas the solid lines represent the regular drawdowns.

3.4.3 *Aggregate performance*

In the rest of this section, we present more backtesting results but in an aggregate way. The goal is not to identify the best strategy, but to see broadly which categories of ETFs tend to have better strategy performance and how the strategies perform compared to the underlying securities on a risk-adjusted based. For this reason, we assume no trading costs. The Sharpe ratio statistics are consistent under leverage, so we do not specify a particular leverage level. In Tables 3.4 and 3.5, we report the annualized averaged category Sharpe ratios for the period Jan 1, 2008 - Sep 30, 2013. The number of ETFs in each category is in the parentheses after the category names in the first column. The average Sharpe ratio of the strategy for each category is reported in the main row in line with the category name, whereas the average Sharpe ratio of the underlying ETFs is immediately below.

Among the 30 categories, 29 categories have higher average Sharpe ratios over their underliers, except for Fixed Income category altogether. In terms of size and style, observations from variance ratios from the first section carry over. Value and large-cap ETFs tend to outperform the growth, mid-cap and small-cap ETFs. For instance, with the style factor fixed, large value ETFs has a total average Sharpe ratio of 0.95 over the entire sample period Jan 1, 2008 - Sep 30, 2013, which is higher than those of mid value ETFs and small value ETFs with Sharpe ratios 0.63 and 0.68 respectively. With the size factor fixed, large (mid or small) value ETFs beat large (mid or small) growth ETFs by Sharpe ratios by about 0.2 points.

By country and region, China and Japan do very well with overall Sharpe ratios at the levels of 1.02 and 1. It is not clear, however, as to what reasons are behind this. For instance, it is not reasonable to argue by society development as China is part of the emerging market and Japan is a developed nation. ETFs based in Latin American (Latam) stocks only have an averaged Sharpe ratio of 0.48. It does appear, however, that ETFs based on firms that operate in a very different time zone from the US have performed well. One may explain China and Japan ETFs have done well because most of the firms these ETFs hold are not traded when the US stock markets are open. Therefore, market participants must estimate these firms' stock prices or the ETFs' prices from what is likely to happen over

Categories	2008-2013	2013	2012	2011	2010	2009	2008
Large (70)	0.83	0.5	0.5	0.74	0.64	0.96	1.61
	0.21	1.07	0.89	-0.19	0.55	1	-0.84
Mid (11)	0.56	0.3	0.26	0.38	0.44	1	1.11
	0.38	1.99	1.06	0	1.05	1.09	-0.94
Small (18)	0.62	0.6	0.34	0.57	0.45	1.13	1.14
	0.33	1.61	1	-0.25	0.96	1.06	-0.89
Large Value (8)	0.95	0.48	0.31	0.93	0.58	0.8	1.87
	0.25	1.82	0.95	0.15	0.64	0.75	-0.87
Large Growth (5)	0.77	0.72	0.7	0.56	0.28	0.76	1.68
	0.35	1.89	1.04	0.09	0.81	1.26	-1.01
Mid Value (5)	0.63	0.26	0.05	0.47	0.47	1.17	1.16
	0.36	1.95	1.09	-0.01	0.91	0.98	-0.85
Mid Growth (3)	0.44	0.32	0.6	0.25	0.4	0.79	0.95
	0.39	2.01	0.99	0.01	1.22	1.26	-1.08
Small Value (7)	0.68	0.5	0.14	0.86	0.31	1.24	1.07
	0.33	1.65	1.16	-0.26	0.94	1.06	-0.94
Small Growth (3)	0.47	0.63	0.42	0.58	0.27	0.54	0.8
	0.45	2.31	0.91	0.1	1.13	1.04	-0.86

Table 3.4: The performance of the strategy and the underlying stock for sub-periods from Jan 1, 2008 to Sep 30, 2013 across various categories. The number of ETFs in each category is in the parentheses after the category names in the first column. The average Sharpe ratio of the strategy for each category is reported in the main row in line with the category name, whereas the average Sharpe ratio of the underlying ETFs is immediately below.

the other side of the world the next day. Such estimation practice can conveniently cause overreactions leading to more significant serial correlation of the ETFs' prices, offering more profitability to our strategy. Commodity, FX and Fixed Income ETFs are pretty much the worst overall performers, with Commodity ETFs having only a Sharpe ratio of 0.28.

We see during the 2008 crisis period, the volatility strategy significantly outperform the underlying ETFs for all categories except fixed income markets. This can be explained by market overreaction and volume imbalance for the period. In addition, this agrees with the fact that our strategy is contrarian in nature. The strategy, however, does not do well when the market is in trending with low volatility. As shown theoretically under the assumption that asset price follows an AR(1) process in the earlier subsection, trends and low volatility are both killers of the profit of the systematic strategy. Empirically, we can see this from various categories in Tables 3.4 and 3.5 during 2012 and 2013, when the underlying ETFs have stellar performances but the strategy has trouble.

Categories	2008 – 2013	2013	2012	2011	2010	2009	2008
International (16)	0.89	1.01	1.2	1.15	1.04	1.39	0.96
	0.23	1.11	0.88	-0.28	0.42	0.97	-1.21
China (4)	1.02	0.9	0.8	-0.14	0.33	1.11	2.87
	0.17	-0.11	0.92	-0.67	0.34	1.22	-0.57
EM (5)	0.81	-0.11	0.79	0.31	0.45	1.38	1.85
	0.15	-0.41	1	-0.76	0.72	1.71	-1.07
EAFE (4)	0.9	0.95	0.61	1.12	0.69	1.23	1.49
	0.14	1.38	0.95	-0.35	0.45	0.98	-1.03
Japan (5)	1	0.73	-0.03	0.39	2.04	2.56	1.62
	0.2	1.23	0.62	-0.43	0.63	0.2	-0.38
Latam (4)	0.31	-0.83	-0.11	0.13	-0.3	0.29	1.04
	0.19	-0.83	0.6	-0.62	0.93	1.82	-0.63
Europe (14)	0.68	0.56	0.77	1.2	0.64	0.81	0.82
	0.12	1.05	1.01	-0.35	0.25	1.02	-1.15
World (11)	0.66	0.4	0.35	0.78	0.65	0.8	1.25
	0.24	1.31	0.92	-0.05	0.58	0.96	-0.92
ConsumerDiscretionaries (5)	0.52	-0.09	0.67	0.34	0.7	0.85	0.84
	0.54	2.43	1.51	0.08	1.12	1.24	-1.09
ConsumerStaples (5)	1.06	0.63	0.9	1.31	1.22	1.41	1.81
	0.56	1.63	1.13	0.64	0.95	0.99	-0.65
Energy (6)	0.63	0.82	0.43	-0.12	0.04	0.88	1.61
	0.21	1.13	0.13	0.13	0.66	0.76	-0.69
Financials (8)	0.9	0.57	-0.03	1.17	0.76	1.56	1.58
	0.15	1.85	1.2	-0.44	0.46	0.61	-0.81
Healthcare (5)	0.81	0.41	0.96	1.11	1.33	0.12	1.03
	0.52	2.56	1.36	0.52	0.21	0.9	-0.72
Industrials (5)	0.51	0.37	0.56	0.5	0.07	0.14	0.7
	0.32	1.94	0.92	-0.11	1.02	0.79	-1.24
Material (5)	0.32	0.34	0.1	0.1	0.16	0.39	0.71
	0.19	0.37	0.55	-0.4	0.81	1.29	-1.07
RealEstate (5)	1.03	-0.71	-0.17	0.71	0.91	2.36	1.8
	0.27	0.28	1.52	0.1	1	0.83	-0.79
Technology (5)	0.68	1.67	0.39	0.4	0.13	0.68	1.41
	0.31	1.08	0.85	-0.06	0.58	1.63	-1.3
Telecommunication (5)	0.78	-0.05	0.39	1.12	0.78	1.53	1.37
	0.29	1.57	0.77	-0.02	0.73	0.04	-0.81
Utility (5)	0.73	-0.08	0.17	1.19	1.4	-0.31	1.49
	0.11	0.95	0.3	0.54	0.22	0.46	-0.74
Commodity (11)	0.28	0.14	0.89	0.25	0.09	0.33	0.54
	-0.02	-0.53	-0.09	-0.2	0.61	0.67	-0.8
Currency (10)	0.46	0.99	1.03	0.65	0.58	1.15	0.23
	0.07	-0.25	0.24	0.01	0.41	0.65	-0.55
FI (14)	0.46	0.3	0.64	0.72	1.35	1.59	0.57
	0.66	-0.47	1.15	1.35	1.14	0.5	0.92

Table 3.5: The performance of the strategy and the underlying stock for sub-periods from Jan 1, 2008 to Sep 30, 2013 across various categories. The number of ETFs in each category is in the parentheses after the category names in the first column. The average Sharpe ratio of the strategy for each category is reported in the main row in line with the category name, whereas the average Sharpe ratio of the underlying ETFs is immediately below.

3.5 Conclusion

In this chapter, we continued our investigation on the market inefficiency induced by serial correlation of asset returns. We took a modeling approach using the Merton's jump-diffusion model, a stationary model and an autoregressive model with time trends to understand the empirical variance ratio behavior observed in Chapter 1 for the period 2006-2013. We found that discretization errors and jumps alone are not capable of generating discretized sample variances which decrease as the return horizons increase. However, a simple autoregressive model, which incorporates the autocorrelation of asset returns, allows such a pattern. We then investigated how the model parameters affect the discretely sampled variance gap using different return frequencies. The stronger the price dynamic mean-reverts, the larger the variance gap. Adding time trends to an AR(1) process reduces the variance gap.

We also proposed a systematic methodology to capture the variance difference sampled at different horizons. It has been backtested over a wide variety of ETFs covering countries, sectors, style and size, asset classes, etc., for the period 2008-2013. We find that the strategy is indeed contrarian and thrives during distressed market conditions. For instance, during 2007-08 financial crisis, the volatility strategy significantly outperform the underlying ETFs for almost all categories in Table 3.5. Although more sophisticated transaction costs and market impact models should be applied to further validate its applicability, we believe that this strategy may have great potential to offer not only similar payoffs provided by over-the-counter derivatives but also extends such opportunities to a much wider collection of securities. The profitability of this strategy provides more evidence against the random walk hypothesis.

3.6 Appendix for Chapter 3

3.6.1 *Drawdown*

The drawdown typically measures the decline from a historical peak of the cumulative profit of a trading strategy or investment. The mathematical formulation of this concept can be defined as follows. Consider a stochastic process X_t for $t \geq 0$, representing the cumulative P&L, with $X_0 = 0$, then the drawdown of this strategy at time T denoted by D_T is defined as

$$D_T = \max 0, \max_{t \in [0,T)} (X_t - X_T). \tag{3.42}$$

The maximum drawdown up to time S is the maximum of the drawdowns over the history $[0, S]$ of the strategy performance. It measures the worst peak to valley loss since the launch of the trading strategy. If we denote it as MD_S, then we can define it as

$$MD_S = \max_{T \in [0,S)} \left(\max_{t \in [0,T)} X_t - X_T \right). \tag{3.43}$$

Therefore, it is easy to see that MD_S is a non-decreasing function in S and it can be constant for an extended time.

Another concept associated with drawdowns is *drawdown duration*, which is essentially the time it takes for any drawdown to recover. Accordingly, the *maximum drawdown duration* is the longest period it takes for a drawdown to recover in the entire history of the strategy.

We have included these concepts either explicitly or implicitly when the variance strategy is backtested and analyzed.

3.6.2 *Proof of Proposition 3.1*

We first consider the Merton jump-diffusion model. The dynamics of this model is given by

$$\frac{dS_t}{S_t} = (r - \lambda m)dt + \sigma dW_t + dJ_t, \tag{3.44}$$

where $J_t = \sum_{i=1}^{N_t}(Y_j - 1)$ and N_t is a Poisson process with rate λ and Y_j is the relative jump size in the stock price. Y_j is log-normally distributed with parameters a and b^2, that is $LN[a, b^2]$, and m is the mean proportional size of jump, i.e. $E(Y_j - 1) = m$. The parameters a, b and m are related to each other via the following equation

$$e^{a+\frac{b^2}{2}} = m + 1. \tag{3.45}$$

When $\lambda = 0$, this model is reduced to the Black-Scholes model.

Consider the sample time interval is [0,T]. We break it down into n pieces of smaller intervals with the ends of intervals denoted by t_i's. Specifically, t_0=0 and $t_n = T$. By applying Ito's Lemma to this jump-diffusion model, and integrate over the interval from t_{i+1} to t_i, we can get

$$\begin{aligned} &\ln S_{i+1} - \ln S_i \\ &= (r - \lambda m - 1/2\sigma^2)(t_{i+1} - t_i) + \sigma\sqrt{t_{i+1} - t_i}N_{i+1} + \sum_{j=1}^{n_j} \ln Y_j \end{aligned} \tag{3.46}$$

where N_{i+1} is iid standard normal random variable coming from the Brownian motion and n_j is the number of jumps in the modeled price during $[t_{i+1}, t_j]$. For simplicity, suppose the t_i's are evenly spaced in the interval [0,T], and define $\Delta t = t_{i+1} - t_i$.

Now to get the discretized variance, we square both sides of the above equation and sum over the index i, for $i = 0$ to $n-1$. The resultant quantity is

$$\begin{aligned} &\sum_{i=0}^{n-1} (\ln S_{i+1} - \ln S_i)^2 \\ &= \sum_{i=0}^{n-1} \left\{ (r - \lambda m - 1/2\sigma^2)^2 \Delta t^2 + \left[\sum_{j=1}^{n_j} \ln Y_j\right]^2 \right. \\ &+ \sigma^2 \Delta t N_{t+1}^2 + 2\sigma(r - \lambda m - 1/2\sigma^2)\Delta t^{1.5} N_{t+1} \\ &\left. + 2(r - \lambda m - 1/2\sigma^2)\Delta t \left[\sum_{j=1}^{n_j} \ln Y_j\right] + 2\sigma\Delta t^{0.5} N_{t+1} \left[\sum_{j=1}^{n_j} \ln Y_j\right] \right\}. \end{aligned}$$

Take the summation into the equation and divide it by T, we get

$$
\begin{aligned}
&\frac{\sum_{i=0}^{n-1} (\ln S_{i+1} - \ln S_i)^2}{T} \\
&= (r - \lambda m - 1/2\sigma^2)^2 \Delta t + \sigma^2 N_{i+1}^2 \\
&+ 2\sigma(r - \lambda m - 1/2\sigma^2)\Delta t^{0.5} N_{i+1} + \left[\sum_{j=1}^{n_j} \ln Y_j\right]^2 / \Delta t \\
&+ 2(r - \lambda m - 1/2\sigma^2)\left[\sum_{j=1}^{n_j} \ln Y_j\right] \\
&+ 2\sigma N_{t+1}\left[\sum_{j=1}^{n_j} \ln Y_j\right] \Delta t^{-0.5}.
\end{aligned}
$$

Since

$$
\begin{aligned}
E(N_i) &= 0 \\
E(N_i^2) &= 1 \\
E\left[\sum_{j=1}^{n_j} \ln Y_j\right] &= a\lambda\Delta t \\
E\left[\sum_{j-1}^{n_j} \ln Y_j\right]^2 &= (a^2 + b^2)\lambda\Delta t + a^2\lambda^2(\Delta t)^2,
\end{aligned}
$$

we have obtained

$$
\begin{aligned}
&E\left[\frac{\sum_{i=0}^{n-1} (\ln S_{i+1} - \ln S_i)^2}{T}\right] \\
&= (r - \lambda m - 1/2\sigma^2)^2 \Delta t + \sigma^2 \\
&+ (a^2 + b^2)\lambda + a^2\lambda^2\Delta t \\
&+ 2(r - \lambda m - 1/2\sigma^2)a\lambda\Delta t \\
&= \sigma^2 + (a^2 + b^2)\lambda \\
&+ \Delta t\left[(r - \lambda m - 1/2\sigma^2)^2 + a^2\lambda^2 + 2(r - \lambda m - 1/2\sigma^2)a\lambda\right].
\end{aligned}
$$

Because this expression is linear in Δt, its derivative in Δt is simply the coefficient of Δt. Therefore,

$$\begin{aligned}F'(\Delta t) &= (r - \lambda m - 1/2\sigma^2)^2 + a^2\lambda^2 + 2(r - \lambda m - 1/2\sigma^2)a\lambda \\ &= (r - m\lambda + a\lambda - \frac{1}{2}\sigma^2)^2.\end{aligned}$$

3.6.3 *Proof of Proposition 3.2*

I The mean of Y_n is zero,

$$EY_n = E[X_n - X_{n-1}] = 0.$$

II Autocovariance defined by Γ_k is

$$\begin{aligned}\Gamma_k &= E\left[Y_n Y_{n-k}\right] \\ &= E\left[(X_n - X_{n-1})(X_{n-k} - X_{n-k-1})\right] \\ &= E\left[X_n X_{n-k} + X_{n-1}X_{n-k-1} - X_{n-1}X_{n-k} - X_n X_{n-k-1}\right] \\ &= 2\gamma_k - \gamma_{k-1} - \gamma_{k+1} \\ &= \frac{\sigma^2 a^{k-1}}{1-a^2}\left[2a - 1 - a^2\right] \\ &= -\frac{\sigma^2 a^{k-1}(1-a)}{1+a} < 0\end{aligned}$$

where $k \geq 1$ and $a \in (0, 1)$.

III Variance defined by Γ_0 is

$$\begin{aligned}\Gamma_0 &= VarY_n \\ &= 2Var(X_n) - 2Cov(X_n, X_{n-1}) \\ &= \frac{2\sigma^2}{1-a^2} - \frac{2\sigma^2 a}{1-a^2} \\ &= \frac{2\sigma^2}{1+a}.\end{aligned}$$

IV Variance of $Y_1 + ... + Y_n$ is

$$\begin{aligned}Var[Y_1 + ... + Y_n] &= Var[X_n - X_0] \\ &= \frac{2\sigma^2(1-a^n)}{1-a^2}.\end{aligned}$$

3.6.4 *Proof of Proposition 3.3*

$$\begin{aligned} Z_k &= Y_1^2 + Y_2^2 + \dots + Y_k^2 - (Y_1 + Y_2 + \dots + Y_k)^2 \\ &= - \sum_{i,j=1, j\neq i}^{k} Y_i Y_j. \end{aligned} \tag{3.47}$$

In particular,

$$\begin{aligned} EZ_k &= -E\left[\sum_{i,j=1,j\neq i}^{k} Y_i Y_j \right] \\ &= 2 \sum_{i,j=1,j>i}^{k} \frac{\sigma^2 a^{j-i-1}}{1-a^2} \left[a^2 - 2a + 1\right] \\ &= 2\sigma^2 \frac{1-a}{1+a} \sum_{l=1}^{k-1} (k-l) a^{l-1} \\ &= 2\sigma^2 \frac{1-a}{1+a} \times \frac{a^k - 1 + k(1-a)}{(1-a)^2} \\ &= \sigma^2 \left(\frac{2k}{1+a} - 2\frac{1-a^k}{(1-a)(1+a)} \right) \geq 0. \end{aligned} \tag{3.48}$$

Now, we compute EZ_k^2. To do so, we make additional assumption on the fourth moment of the noise w_n such that

$$Ew_i w_j w_l w_m = \begin{cases} \eta\sigma^4, \text{if } i = j = l = m \\ \sigma^4, \text{if } i = j,\ l = m \text{ or } i = l,\ j = m \text{ or } i = m,\ j = l \\ 0, \text{otherwise.} \end{cases}$$

If we represent Y_n by

$$Y_n = \sum_{i=0}^{\infty} \alpha_i w_{n-i}, \tag{3.49}$$

where $\alpha_0 = 1$ and $\alpha_i = (a-1)a^{i-1}$ for all $i \geq 1$, then, for $h, p, q \geq 0$, we have

$$\begin{aligned} &EY_n Y_{n+h} Y_{n+h+p} Y_{n+h+p+q} \\ &= (\eta - 3)\sigma^4 \sum_{t=0}^{\infty} \alpha_t \alpha_{t+h} \alpha_{t+h+p} \alpha_{t+h+p+q} + \Gamma_h \Gamma_q + \Gamma_{h+p} \Gamma_{p+q} + \Gamma_{h+p+q} \Gamma_p. \end{aligned} \tag{3.50}$$

In particular, we specify as follows ϕ_1, ϕ_2, ϕ_3 and ϕ_4, which will be used as parts of the expression of EZ_k^2.

I Define $\phi_1(p) = EY_n^2 Y_{n+p}^2$ for $p > 0$, then

$$\phi_1(p) = \Gamma_0^2 + 2\Gamma_p^2 + (\eta - 3)\sigma^4 \sum_{t=0}^{\infty} \alpha_t^2 \alpha_{t+p}^2 \tag{3.51}$$

$$= \Gamma_0^2 + 2\Gamma_p^2$$

$$+ (\eta - 3)\sigma^4 \left[(a-1)^2 a^{2p-2} + (a-1)^4 a^{2p-4} \sum_{t=1}^{\infty} a^{4t} \right]$$

$$= \frac{4\sigma^4}{(1+a)^2} + 2\frac{\sigma^4 a^{2p-2}(a-1)^2}{(1+a)^2}$$

$$+ (\eta - 3)\sigma^4 \left[(a-1)^2 a^{2p-2} + (a-1)^4 \frac{a^{2p}}{1-a^4} \right].$$

II Define $\phi_2(p,q) = EY_n^2 Y_{n+p} Y_{n+p+q}$ for $p > 0$ and $q > 0$, then

$$\phi_2(p,q) = \Gamma_0 \Gamma_q + 2\Gamma_{p+q}\Gamma_p + (\eta - 3)\sigma^4 \sum_{t=0}^{\infty} \alpha_t^2 \alpha_{t+p} \alpha_{t+p+q} \tag{3.52}$$

$$= \Gamma_0 \Gamma_q + 2\Gamma_{p+q}\Gamma_p$$

$$+ (\eta - 3)\sigma^4 \left[(a-1)^2 a^{2p+q-2} + (a-1)^4 a^{2p+q-4} \sum_{t=1}^{\infty} a^{4t} \right]$$

$$= \frac{2\sigma^4 a^{q-1}(a-1)}{(1+a)^2} + 2\frac{\sigma^4 a^{2p+q-2}(a-1)^2}{(1+a)^2}$$

$$+ (\eta - 3)\sigma^4 \left[(a-1)^2 a^{2p+q-2} + (a-1)^4 \frac{a^{2p+q}}{1-a^4} \right].$$

III Define $\phi_3(h,q) = EY_n Y_{n+h}^2 Y_{n+h+q}$ for $h > 0$ and $q > 0$, then

$$\phi_3(h,q) = \Gamma_0 \Gamma_{h+q} + 2\Gamma_h \Gamma_q + (\eta - 3)\sigma^4 \sum_{t=0}^{\infty} \alpha_t \alpha_{t+h}^2 \alpha_{t+h+q} \tag{3.53}$$

$$= \Gamma_0 \Gamma_{h+q} + 2\Gamma_h \Gamma_q$$

$$+ (\eta - 3)\sigma^4 \left[(a-1)^3 a^{3h+q-3} + (a-1)^4 a^{3h+q-4} \sum_{t=1}^{\infty} a^{4t} \right]$$

$$= \frac{2\sigma^4 a^{h+q-1}(a-1)}{(1+a)^2} + 2\frac{\sigma^4 a^{h+q-2}(a-1)^2}{(1+a)^2}$$

$$+ (\eta - 3)\sigma^4 \left[(a-1)^3 a^{3h+q-3} + (a-1)^4 \frac{a^{3h+q}}{1-a^4} \right].$$

IV Define $\phi_4(h,p,q) = EY_nY_{n+h}Y_{n+h+p}Y_{n+h+p+q}$ for $h > 0$, $p > 0$ and $q > 0$, then

$$\begin{aligned}
\phi_4(h,p,q) &= \Gamma_h\Gamma_q + \Gamma_{h+p}\Gamma_{p+q} + \Gamma_{h+p+q}\Gamma_p \\
&+ (\eta-3)\sigma^4\sum_{t=0}^{\infty}\alpha_t\alpha_{t+h}\alpha_{t+h+p}\alpha_{t+h+p+q} \\
&= \Gamma_h\Gamma_q + \Gamma_{h+p}\Gamma_{p+q} + \Gamma_{h+p+q}\Gamma_p \\
&+ (\eta-3)\sigma^4\left[(a-1)^3a^{3h+2p+q-3} + (a-1)^4a^{3h+2p+q-4}\sum_{t=1}^{\infty}a^{4t}\right] \\
&= \frac{\sigma^4a^{h+q-2}(a-1)^2}{(1+a)^2} + 2\frac{\sigma^4a^{h+2p+q-2}(a-1)^2}{(1+a)^2} \\
&+ (\eta-3)\sigma^4\left[(a-1)^3a^{3h+2p+q-3} + (a-1)^4\frac{a^{3h+2p+q}}{1-a^4}\right].
\end{aligned} \tag{3.54}$$

Thus,

$$\begin{aligned}
EZ_k^2 &= E\sum_{\substack{i,j=1\\ i\neq j}}^{k}Y_iY_j\sum_{\substack{l,m=1\\ l\neq m}}^{k}Y_lY_m \\
&= 4\sum_{\substack{1\leq i<j\leq k\\ 1\leq l<m\leq k}}EY_iY_jY_lY_m \\
&= 4\sum_{\substack{1\leq i<j\leq k\\ 1\leq i<m\leq k}}EY_i^2Y_jY_m + 8\sum_{\substack{1\leq i<j\leq k\\ 1\leq l<m\leq k}}^{i<l}EY_iY_jY_lY_m \\
&= 4\sum_{1\leq i<j\leq k}EY_i^2Y_j^2 + 8\sum_{1\leq i<j<m\leq k}EY_i^2Y_jY_m + 8\sum_{\substack{1\leq i<j\leq k\\ 1\leq l<m\leq k}}^{i<l}EY_iY_jY_lY_m \\
&= 4\sum_{1\leq i<j\leq k}EY_i^2Y_j^2 + 8\sum_{1\leq i<j<m\leq k}EY_i^2Y_jY_m \\
&+ 8\sum_{1\leq i<j<m\leq k}EY_iY_j^2Y_m + 16\sum_{1\leq i<j<l<m\leq k}EY_iY_jY_lY_m \\
&= 4\sum_{i=1}^{k-1}\sum_{j=i+1}^{k}\phi_1(j-i) + 8\sum_{i=1}^{k-2}\sum_{j=i+1}^{k-1}\sum_{m=j+1}^{k}\phi_2(j-i,m-j) \\
&+ 8\sum_{i=1}^{k-2}\sum_{j=i+1}^{k-1}\sum_{m=j+1}^{k}\phi_3(j-i,m-j)
\end{aligned}$$

$$+16\sum_{i=1}^{k-3}\sum_{j=i+1}^{k-2}\sum_{l=j+1}^{k-1}\sum_{m=l+1}^{k}\phi_4(j-i,l-j,m-l). \tag{3.55}$$

We define

$$f(\sigma,k,a) = EZ_k = \sigma^2\left(\frac{2k}{1+a} - 2\frac{1-a^k}{(1-a)(1+a)}\right) \tag{3.56}$$

$$s(\sigma,k,a,\eta)^2 = EZ_k^2 - E^2Z_k, \tag{3.57}$$

and investigate further their properties with $a \in (0,1)$, $k = 1,2,3,...$ and $\sigma > 0$. Since the expression of $s(\sigma,k,a,\eta)$ is rather complex, we understand its behavior graphically. For $f(\sigma,k,a)$, however, we will establish four basic properties using its closed form solution.

I By L'Hopital's rule, it is easy to see

$$\begin{aligned}\lim_{a\to1^-} f(\sigma,k,a) &= \lim_{a\to1^-}\sigma^2\left(\frac{2k}{1+a} - 2\frac{1-a^k}{(1-a)(1+a)}\right)\\ &= \sigma^2\left(k - \lim_{a\to1^-} 2\frac{1-a^k}{(1-a)(1+a)}\right)\\ &= \sigma^2 k - 2\sigma^2\lim_{a\to1^-}\frac{-ka^{k-1}}{-2a}\\ &= 0,\end{aligned} \tag{3.58}$$

which is consistent with our understanding of the case where X_n follows a random walk.

II $f(\sigma,k,a)$ is increasing in k for all $k = 1,2,3,...$, because

$$\begin{aligned}&f(\sigma,k+1,a) - f(\sigma,k,a)\\ &= \sigma^2\left(\frac{2k+2}{1+a} - 2\frac{1-a^{k+1}}{(1-a)(1+a)}\right) - \sigma^2\left(\frac{2k}{1+a} - 2\frac{1-a^k}{(1-a)(1+a)}\right)\\ &= \frac{2\sigma^2}{1+a} - \frac{2\sigma^2a^k(1-a)}{(1-a)(1+a)}\\ &= \frac{2\sigma^2(1-a^k)}{1+a} > 0\end{aligned} \tag{3.59}$$

for all k given $a \in (0,1)$.

III $f(\sigma,k,a)$ decreases in a for $k \geq 2$. When $k = 1$, $f(\sigma,a,k) = 0$ for all $a \in (0,1)$ and $\sigma > 0$. We take the partial derivative of f with respect to a.

$$f_a(\sigma,k,a) \tag{3.60}$$

$$= -2\sigma^2 \frac{k(1-a)\left[1-a-a^{k-1}(1+a)\right] + 2a(1-a^k)}{(1+a)^2(1-a)^2}.$$

Note, for $k \geq 2$, $a \in (0,1)$, the term $2a(1-a^k)$ is always positive. So, we focus on the term $1-a-a^{k-1}(1+a)$. Since $a \in (0,1)$, a^{k-1} is decreasing in k, the whole term increases with k. When $k=2$, we have $1-a^2 > 0$, thus, this term is positive for all $k = 2, 3, ..., n-1$. Hence, $f_a(\sigma, k, a) < 0$.

IV Since $f(\sigma, k, a)$ is always positive, it increases as σ increases.

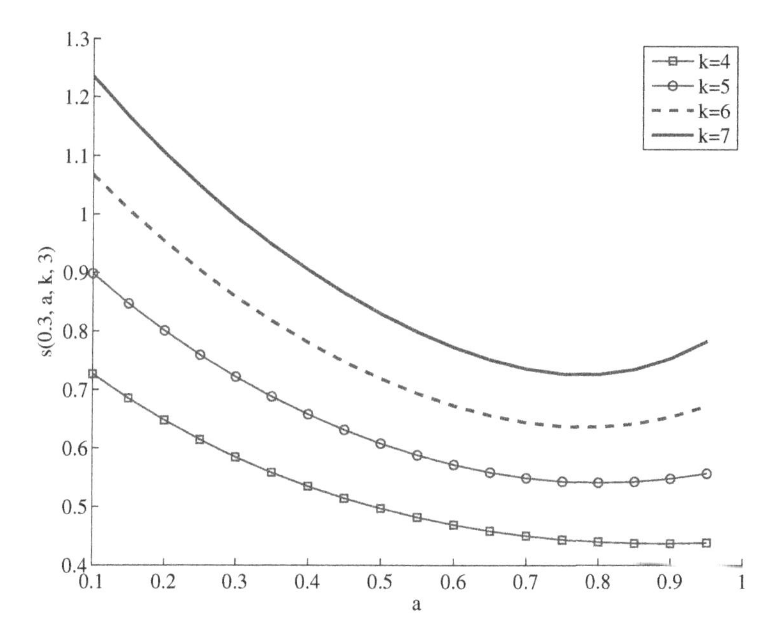

Fig. 3.14: $s(0.3, k, a, 3)$ for $k = 4, 5, 6$ and 7 across $a \in (0,1)$, where we have indicated $\sigma = 0.3$ and $\eta = 3$. For k being small natural numbers, the function is monotone decreasing. As k increases in size, the curve becomes convex, which first decreases and then increases as a increases from 0 to 1.

Due to the complexity of the form of $s(\sigma, k, a, \eta)$, we investigate its properties graphically. We plot $s(\sigma, k, a, \eta)$ in the following figures. Figure 3.14 plots $s(0.3, k, a, 3)$ for $k = 4, 5, 6$ and 7 across $a \in (0,1)$, where we have indicated $\sigma = 0.3$ and $\eta = 3$. For k being small natural numbers, the function is monotone decreasing. As k increases in size, the curve becomes convex, which first decreases and then increases as a increases from 0 to 1.

Figure 3.15 plots $s(\sigma, 5, a, 3)$ for $\sigma = 0.1, 0.2, 0.3$ and 0.4 across $a \in (0, 1)$, where we have indicated $k = 5$ and $\eta = 3$. The function is convex in a. As σ increases, the function curve shifts upward and its curvature increases.

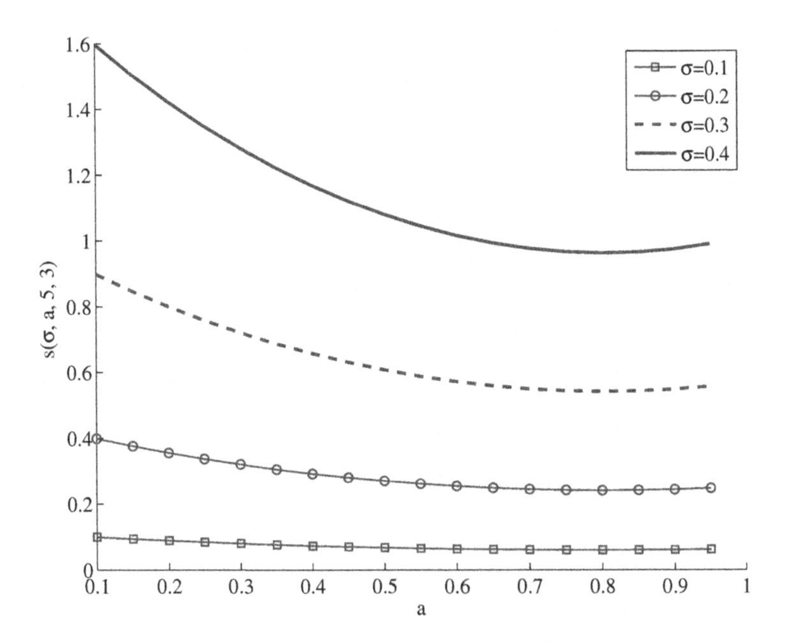

Fig. 3.15: $s(\sigma, 5, a, 3)$ for $\sigma = 0.1, 0.2, 0.3$ and 0.4 across $a \in (0, 1)$, where we have indicated $k = 5$ and $\eta = 3$. The function is convex in a. As σ increases, the function curve shifts upward and its curvature increases.

Figure 3.16 plots $s(0.3, 5, a, \eta)$ for $\eta = 1, 2, 3, 4$ and 5 across $a \in (0, 1)$, where we have indicated $k = 5$ and $\sigma = 0.3$. η specifies the kurtosis of the noise of the AR(1) process. If the noise follows a normal distribution, then $\eta = 3$. If $\eta > 3$, the noises are of a fat-tailed distribution. The function is convex in a. As η increases, the function curve shifts downward and its curvature increases.

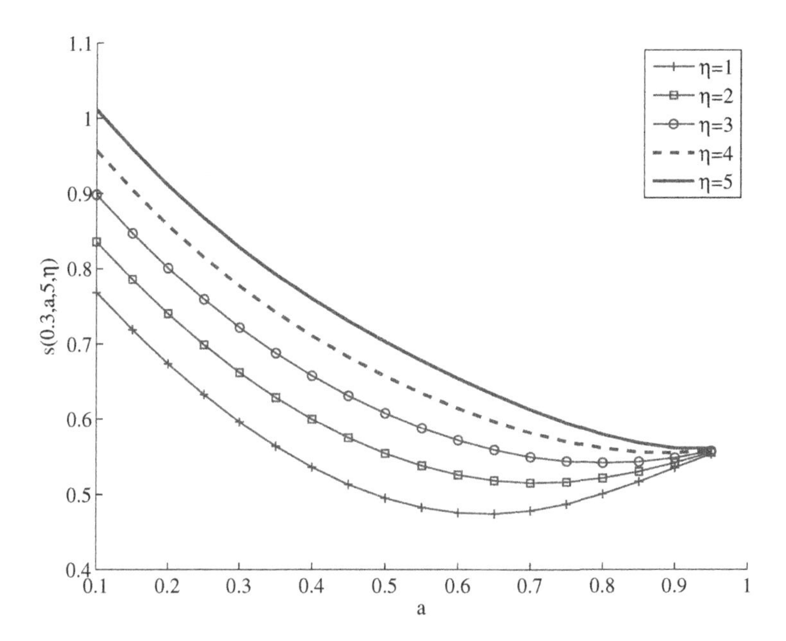

Fig. 3.16: $s(0.3, 5, a, \eta)$ for $\eta = 1, 2, 3, 4$ and 5 across $a \in (0, 1)$, where we have indicated $k = 5$ and $\sigma = 0.3$. The function is convex in a. As η increases, the function curve shifts downward and its curvature increases.

3.6.5 *Proof of Proposition 3.4*

We compute the following quantities.

$$\begin{aligned}
ES_n &= d + cn + \mu \\
E[S_n S_{n-k}] &= E\left[(d + cn + X_n)(d + c(n-k) + X_{n-k})\right] \\
&= (d + cn)(d + c(n-k)) + (d + cn)EX_{n-k} \\
&+ (d + c(n-k))EX_n + EX_n X_{n-k} \\
&= d^2 + cd(2n-k) + c^2 n(n-k) + [2d + c(2n-k)]\mu + \gamma_k + \mu^2
\end{aligned}$$

for all n and $k = 0, 1, 2, ...$, where μ and γ_k are the same as defined for the stationary model.

Denote the one step difference of S_n by Y_n, that is,

$$Y_n = S_n - S_{n-1}. \tag{3.61}$$

Then,

$$
\begin{aligned}
E\left[Y_n Y_{n-k}\right] &= E\left[(S_n - S_{n-1})(S_{n-k} - S_{n-k-1})\right] \\
&= E\left[S_n S_{n-k} + S_{n-1}S_{n-k-1} - S_{n-1}S_{n-k} - S_n S_{n-k-1}\right] \\
&= d^2 + cd(2n-k) + c^2 n(n-k) + [2d + c(2n-k)]\mu + \gamma_k + \mu^2 \\
&+ d^2 + cd(2n-k-2) + c^2(n-1)(n-k-1) \\
&+ [2d + c(2n-k-2)]\mu + \gamma_k + \mu^2 \\
&- d^2 + cd(2n-k-1) + c^2(n-1)(n-k) \\
&+ [2d + c(2n-k-1)]\mu + \gamma_{k-1} + \mu^2 \\
&- d^2 + cd(2n-k-1) + c^2 n(n-k-1) \\
&+ [2d + c(2n-k-1)]\mu + \gamma_{k+1} + \mu^2 \\
&= c^2 + 2\gamma_k - \gamma_{k-1} - \gamma_{k+1} \\
&= c^2 + \frac{\sigma^2 a^{k-1}}{1-a^2}\left[2a - 1 - a^2\right].
\end{aligned} \tag{3.62}
$$

Therefore,

$$
\begin{aligned}
Z_k &= (S_k - S_{k-1})^2 + (S_{k-1} - S_{k-2})^2 \\
&+ (S_1 - S_0)^2 - \ldots - (S_k - S_0)^2 \\
&= Y_k^2 + Y_{k-1}^2 + \ldots + Y_1^2 - (Y_k + Y_{k-1} + \ldots + Y_1)^2 \\
&= - \sum_{i,j=1, j \neq i}^{k} Y_i Y_j.
\end{aligned} \tag{3.63}
$$

In particular,

$$
\begin{aligned}
EZ_k &= E\left[- \sum_{i,j=1, j \neq i}^{k} Y_i Y_j\right] \\
&= -k(k-1)c^2 + 2 \sum_{i,j=1, j<i}^{k} \frac{\sigma^2 a^{i-j-1}}{1-a^2}\left[a^2 - 2a + 1\right] \\
&= f(\sigma, k, a) - k(k-1)c^2.
\end{aligned} \tag{3.64}
$$

Denote $g(\sigma, k, a, c) := EZ_k = f(\sigma, k, a) - k(k-1)c^2$, then

$$
g(\sigma, k, a, c) = \sigma^2 \left(\frac{2k}{1+a} - 2\frac{1-a^k}{(1-a)(1+a)}\right) - k(k-1)c^2. \tag{3.65}
$$

3.6.6 *The collection of ETFs used in this chapter*

Large					Mid	Small		Large Value	Large Growth
IWD	VV	PIN	EWY	AXHE	IWS	IWN	EWGS	DIA	EFG
IVE	DWM	INDY	ILF	IXJ	IJJ	IJS	SCZ	DLN	EGRW
DIA	DOL	EWJ	XLY	XLI	VOE	VBR	EEMS	DOO	IVW
VTV	FXI	DXJ	RXI	EXI	DON	DLS	EWX	EFV	IWF
VIG	MCHI	ITF	EMDI	MXI	DIM	DFE	SCIF	EVAL	QQQ
DLN	FCHI	EWA	XLP	XLK	IWP	DES	SMIN	IVE	VUG
DOO	ERUS	EWO	ECON	IXN	IJK	DGS	DFJ	IWD	
EFV	RBL	EWC	KXI	AXIT	VOT	IWO	SCJ	VIG	
EVAL	EWZ	EWQ	AXSL	IXP	IWR	IJT	DXJS	VTV	
QQQ	EWU	EWH	XLE	AXTE	MDY	VBK	PSCD		
IWF	VGK	EIS	IXC	XLU	VO	IWM	PSCC		
IVW	FEU	EWI	AXEN	JXI	IVOO	SLY	PSCE		
VUG	EWG	ENZL	EMEY	EMIF	BRAZ	VB	PSCF		
EFG	EFA	EWS	XLF	IPU		HAO	PSCH		
EGRW	VWO	EWP	VFH	AXUT		ECNS	PSCI		
GXC	EEM	EWD	EUFN			RSXJ	PSCM		
YAO	BKF	EWL	IXG			BRF	ROOF		
IWB	INP	EWW	AXFN			EWZS	PSCT		
SPY	INDA	EWT	XLV			EWUS	PSCU		

Table 3.6: The ETFs of the following categories: Large, Mid, Small, Large Value and Large Growth.

Mid Value	Mid Growth	Small Value	Small Growth	International		China	EM	EAFE	Japan
IWS	IWP	IWN	IWO	AXEN	IRV	FCHI	BKF	EFA	DXJ
IJJ	IJK	IJS	IJT	AXFN	AXMT	FXI	DGS	SCZ	EWJ
VOE	VOT	VBR	VBK	AXHE	RWX	GXC	ECON	EFG	DFJ
DON		DLS		AXIT	VNQI	HAO	EEM	EFV	SCJ
DIM		DFE		AXSL	IPK	ECNS	EMDI		DXJS
		DES		AXTE	DOO	MCHI	EMEY		ITF
		DGS		AXUT	IST	YAO	EDIV		
				DOL			EMIF		
				DIM			EEMS		
				DLS			EWX		
				DWM			VWO		
				IPU			EMFN		
				IPD			EGRW		
				IPS			EMMT		
				IPW			EVAL		
				IPF					
				IRY					
				IPN					
				AXID					

Table 3.7: The ETFs of the following categories: Mid Value, Mid Growth, Small Value, Small Growth, International, China, EM, EAFE and Japan.

Latam	Europe		World	Consumer Discretionaries	Consumers Staples	Energy	Financials	Healthcare
EWW	DFE	EPOL	EXI	EMDI	AXSL	AXEN	AXFN	AXHE
EWZ	EUFN	TUR	IXC	VCR	ECON	EMEY	EUFN	IXJ
BRAZ	EWD	VGK	IXG	IYC	KXI	IXC	IXG	XLV
BRF	EWG		IXJ	PSCD	VDC	IYE	VFH	VHT
EWZS	EWI		IXN	IPD	IYK	XLE	XLF	IYH
ILF	EWL		IXP	RXI	PSCC	IEZ	IYF	PSCH
ECH	EWO		JXI	XLY	IPS	VDE	IAT	IRY
EPU	EWP		MXI		XLP	PSCE	IAI	
AND	EWQ		RXI			IPW	IAK	
	EWU		QQQ				PSCF	
	FEU		RWO				IPF	
	EWUS						EMFN	
	EWGS							
	EWK							
	EDEN							
	EFNL							
	EIRL							
	EWN							
	ENOR							

Table 3.8: The ETFs of the following categories: Latam, Europe, World, Consumer Staples, Consumer Discretionaries, Energy, Financials and Healthcare.

Industrials	Materials	REIT	Technology	Telecom	Utility	Commodity	Currency	FI
EXI	MXI	VNQ	AXIT	AXTE	AXUT	GLD	FXA	LQD
XLI	XLB	IYR	IXN	IXP	EMIF	SLV	FXC	HYG
IYJ	VAW	RWX	XLK	VOX	IPU	USO	FXF	BSV
VIS	IYM	RWR	VGT	IYZ	JXI	UNG	FXE	SHY
PSCI	PSCM	RWO	IYW	IST	XLU	DBB	FXY	IEI
IPN	IRV	ROOF	PSCT	XTL	VPU	DBE	DBV	IEF
AXID	EMMT	VNQI	IPK	VGT	IDU	DBP	FXB	EMB
	AXMT				PSCU	DBO	FXS	BIV
						DBA	CNY	MUB
						DBC	INR	TLT
						GSP	FXSG	BWX
								BIL
								MBB
								AGZ

Table 3.9: The ETFs of the following categories: Industrials, Materials, REIT, Technology, Telecom, Utility, Commodity, Currency and FI.

Bibliography

Almgren, R. and Chriss, N. (2001). Optimal execution of portfolio transactions. *Journal of Risk* **3**, pp. 5–40.

Anderson, R. M., Eom, K. S., Hahn, S. B., and Park, J.-H. (2012). Sources of stock return autocorrelation. `http://emlab.berkeley.edu/~anderson/Sources-042212.pdf`.

Avramov, D., Chordia, T., and Goyal, A. (2006). Liquidity and autocorrelations in individual stock returns. *The Journal of Finance* **61**, 5, pp. 2365–2394.

Azzarello, S. (2014). Energy price spread: Natural gas vs. crude oil in the us. `http://www.cmegroup.com/education/files/energy-price-spread-natural-gas-vs-crude-oil-in-the-us.pdf`.

Boudoukh, J., Richardson, M. P., and Whitelaw, R. (1994). A tale of three schools: Insights on autocorrelations of short-horizon stock returns. *Review of Financial Studies* **7**, 3, pp. 539–573.

Broadie, M. and Jain, A. (2008). The effect of jumps and discrete sampling on volatility and variance swaps. *International Journal of Theoretical and Applied Finance* **11**, 8, pp. 761–797

Campbell, J. Y., Grossman, S. J., and Wang, J. (1993). Trading volume and serial correlation in stock returns. *The Quarterly Journal of Economics* **108**, 4, pp. 905–939.

Carr, P., Stanley, M., and Madan, D. (1998). Towards a theory of volatility trading, in *Reprinted in Option Pricing, Interest Rates, and Risk Management, Musiella, Jouini, Cvitanic* (University Press), pp. 417–427.

Cheung, K.-C. and Coutts, J. A. (2001). A note on weak form market efficiency in security prices: Evidence from the hong kong stock exchange. *Applied Economics Letters* **8**, 6, pp. 407–410.

Chiang, S.-M., Lee, Y.-H., Su, H.-M., and Tzou, Y.-P. (2010). Efficiency tests of foreign exchange markets for four asian countries, *Research in International Business and Finance* **24**, 3, pp. 284–294.

Choi, J., Salandro, D., and Shastri, K. (1988). On the estimation of bid-ask spreads: Theory and evidence. *Journal of Financial and Quantitative Analysis* **23**, 2, pp. 219–230.

Conover, C. M., Jensen, G. R., Johnson, R. R., and Mercer, J. M. (2008). Sector rotation and monetary conditions, *The Journal of Investing* **17**, 1, pp. 34–46.

Dezelan, S. (2000). Efficiency of the slovenian equity market. *Economic and Business Review* **2**, 1, pp. 61–83.

Dickey, D. and Fuller, W. A. (1979). Distribution of the estimators for autoregressive time series with a unit root. *Journal of the American Statistical Association* **74**, 366a, pp. 427–431.

Fama, E. F. and French, K. R. (1992). The cross-section of expected stock returns. *The Journal of Finance* **47**, 2, pp. 427–465.

Fama, E. F. and French, K. R. (1993). Common risk factors in the returns on stocks and bonds. *Journal of Financial Economics* **33**, 1, pp. 3–56.

Fleming, M. J. and Remolona, E. M. (1997). Price formation and liquidity in the us treasury market: evidence from intraday patterns around announcements. *FRB of New York Staff Report* **27**.

Garleanu, N. and Pedersen, L. H. (2013). Dynamic trading with predictable returns and transaction costs. *The Journal of Finance* **68**, 6, pp. 2309–2340.

Geman, H. (2007). Mean reversion versus random walk in oil and natural gas prices, in *Advances in Mathematical Finance* (Springer), pp. 219–228.

Glosten, L. R. and Milgrom, P. R. (1985). Bid, ask and transaction prices in a specialist market with heterogeneously informed traders. *Journal of Financial Economics* **14**, 1, pp. 71–100.

Grieb, T. and Reyes, M. G. (1999). Random walk tests for latin american equity indexes and individual firms. *Journal of Financial Research* **22**, 4, pp. 371–383.

Groenewold, N. and Ariff, M. (2002a). The effects of de-regulation on share market efficiency in the asia-pacific. *International Economic Journal* **12**, 4, pp. 23–47.

Groenewold, N. and Ariff, M. (2002b). Weak-form efficiency in the czech equity market. *Politicka Ekonomie* **50**, 3, pp. 377–389.

Han, J. (1991). *The risk and return characteristics of real estate investment trusts* (MIT).

Harris, L. (1990). Statistical properties of the roll serial covariance bid/ask spread estimator. *The Journal of Finance* **45**, 2, pp. 579–590.

Hong, Y., Lin, H., and Wu, C. (2012). Are corporate bond market returns predictable? *Journal of Banking & Finance* **36**, 8, pp. 2216–2232.

Huang, B.-N. (1995). Do Asian stock market prices follow random walks? Evidence from the variance ratio test, *Applied Financial Economics* **5**, 4, pp. 251–256.

Hull, J. (2009). *Options, futures and other derivatives.* (Pearson education).

Jensen, M. C. (1967). The performance of mutual funds in the period 1945-1964. *The Journal of Finance* **23**, 2, pp. 389–416.

Jensen, M. C. (1978). Some anomalous evidence regarding market efficiency. *The Journal of Financial Economics* **6**, 2, pp. 95–101.

Jirasakuldech, B. and Knight, J. R. (2005). Efficiency in the market for reits: Further evidence. *Journal of Real Estate Portfolio Management* **11**, 2, pp.

123–132.

Kuhle, J. L. and Alvayay, J. R. (2000). The efficiency of equity REIT prices. *Journal of Real Estate Portfolio Management* **6**, 4, pp. 349–354.

Lakonishok, J. and Lee, I. (2001). Are insider trades informative? *Review of Financial Studies* **14**, 1, pp. 79–11.

LeBaron, B. (1992). Some relations between volatility and serial correlations in stock market returns, *The Journal of Business* **65**, 2, pp. 199–219.

Lee, C. F., Chen, G.-M., and Rui, O. M. (2001). Stock reruns and volatility on china's stock markets. *The Journal of Financial Research* **24**, 4, pp. 523–544.

LeRoy, S. F. (1973). Risk aversion and the martingale property of stock prices. *International Economic Review* **14**, 2, pp. 436–446.

Ljung, G. M. and Box, G. E. (1978). On a measure of lack of fit in time series models. *Biometrika* **65**, 2, pp. 297–303.

Lo, A. W. and MacKinlay, A. C. (1988). Stock market prices do not follow random walks: Evidence from a simple specification test. *Review of Financial Studies* **1**, 1, pp. 41–66.

Lo, A. W. and MacKinlay, A. C. (1990). When are contrarian profits due to stock market overreaction?. *Review of Financial Studies* **3**, 2, pp. 175–205.

Malkiel, B. G. and Fama, E. F. (1970). Efficient capital markets: a review of theory and empirical work. *The Journal of Finance* **25**, 2, pp. 383–417.

Mech, T. (1993). Portfolio return autocorrelation. *Journal of Financial Economics* **34**, 3, pp. 307–344.

Nick, S. (2013). Price formation and intertemporal arbitrage within a low-liquidity framework: Empirical evidence from european natural gas markets. *Energiewirtschaftliches Institut an der Universitt zu Koeln* **14**.

Poon, S.-H. (1996). Persistence and mean reversion in UK stock returns. *European Financial Management* **2**, 2, pp. 169–196.

Porta, R. L., Lakonishok, J., Shleifer, A., and Vishny, R. (1997). Good news for value stocks: Further evidence on market efficiency. *The Journal of Finance* **52**, 2, pp. 859–874.

Roley, V. V. (1982). Weekly money supply announcements and the volatility of short-term interest rates. *Economic Review* **67**, pp. 3–15.

Roll, R. (1984). A simple implicit measure of the effective bid-ask spread in an efficient market. *The Journal of Finance* **39**, 4, pp. 1127–1139.

Sentana, E. and Wadhwani, S. (1992). Feedback traders and stock return autocorrelations: evidence from a century of daily data. *The Economic Journal* **102**, 411, pp. 415–425.

Smith, G. (2002). Tests of the random walk hypothesis for london gold prices. *Applied Economics Letters* **9**, 10, pp. 671–674.

Tucker, M. and Laipply, S. (2012). High yield etf behavior in stressed markets. `http://www.blackrock.com/corporate/en-hk/literature/whitepaper/high-yield-etf-behavior-in-stressed-markets.pdf`.

Tucker, M. and Laipply, S. (2013). Bond market price discovery: Clarity through the lens of an exchange. *The Journal of Portfolio Management* **39**, 2, pp. 49–62.

Wang, K., Erickson, J., Gau, G., and Chan, S. H. (1995). Market microstructure and real estate returns. *Real Estate Economics* **23**, 1, pp. 85–100.

Wang, Y. and Liu, L. (2010). Is WTI crude oil market becoming weakly efficient over time? New evidence from multi scale analysis based on detrained fluctuation analysis. *Energy Economics* **32**, 5, pp. 987–992.

Worthington, A. C. and Higgs, H. (2004). Random walks and market efficiency in european equity markets. *Global Journal of Finance and Economics* **1**, 1, pp. 59–78.

Index